M

MONOGRAPHS IN GEOLOGY AND
PALEONTOLOGY · 1
EDITED BY ALFRED G. FISCHER

ELECTRON MICROGRAPHS
OF LIMESTONES
AND THEIR NANNOFOSSILS

18837

Electron micrograph of Eocene limestone with nanno-fossils; Crescent or Metchosin formation, Olympic peninsula, Washington. The star-shaped object is a section through *Discoaster barbadiensis*; each of its nine rays is a calcite crystal, showing cleavage. The mosaic of grains below represents a tangential section through the wall of *Thoracosphaera*. The many-sutured grains to the left of the discoaster are coccoliths. The highstanding grains surrounded by dark veils (shreds of replica) are clastic grains of quartz or silicates. Scale bar 5 microns. Photo Garrison.

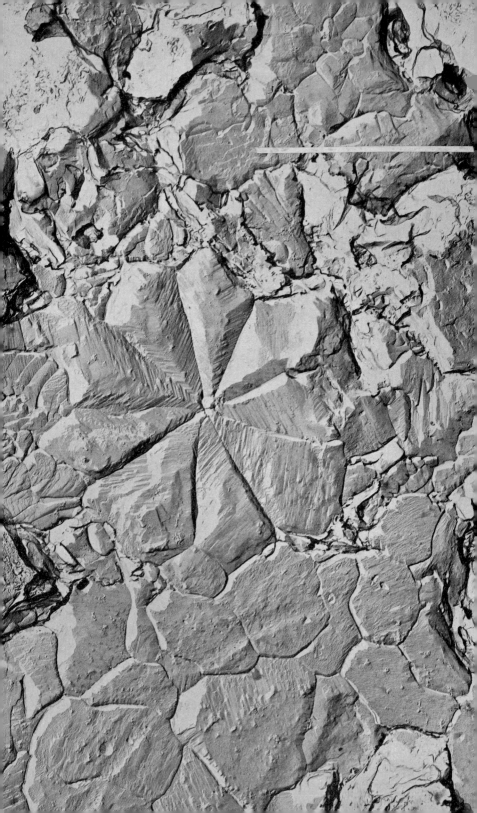

ELECTRON MICROGRAPHS
OF LIMESTONES
AND THEIR NANNOFOSSILS

ALFRED G. FISCHER

SUSUMU HONJO

ROBERT E. GARRISON

PRINCETON, NEW JERSEY
PRINCETON UNIVERSITY PRESS
1967

Dedicated to

BRUNO SANDER

PREFACE

Somewhat over a century ago, the development of the thin section and the petrographic microscope extended the study of rocks and their constituent grains into the microscopic realm—to magnifications ranging from tens to hundreds and, in favorable cases, to around a thousand diameters. In recent decades, the development of the electron microscope has extended the range of possible investigation by several orders of magnitude. To date, the electron microscope has been used mainly in the study of isolated particles: clay crystals, surfaces of sand grains, and ultramicroscopic fossils such as coccoliths. However, a number of investigators, among them Folk and Weaver (1952), Grunau and Studer (1956), Seeliger (1956), Grunau (1959), Grégoire and Monty (1963), Shoji and Folk (1964), Farinacci (1964), Flügel and Fenninger (1966), Harvey (1966), Flügel (1967), and the present authors, have used it to study whole rocks.

Our study was designed as a survey of fine-grained limestones in order to learn something about the fabric of these rocks, i.e., the sizes, shapes, and arrangement of their constituent grains, as well as about the nature of these grains, thus to gain new insight into the different kinds of limestones and into the processes which have formed them. We initially planned to use a combination of optical and electron microscopy; we developed methods for cutting comparatively thin sections for optical investigation, and made some progress in the optical study of peel replicas (Honjo and Fischer, 1965b). But we found comparatively low-power electron micrographs of polished and etched sections of limestones to yield so much more information (Honjo and Fischer, 1965a) that we concentrated largely on this technique.

The micrographs reproduced in this book are a sampling of some hundreds of pictures of limestones ranging in age from Cambrian to Recent, in origin from intertidal to deep water, in alteration from virtually unconsolidated to tightly cemented to mildly metamorphosed, and in structural attitude from flat lying to highly folded. Our coverage is, however, very uneven, for we found the nanno-fossiliferous[1] deeper-water pelagic limestones of Mesozoic and Cenozoic age so interesting that we spent far more effort on them than on Paleozoic rocks or limestones of other facies. This imbalance may be partly offset by the fact that a fair number of excellent electron micrographs of Paleozoic limestones have been published by Grégoire and Monty (1963), by Shoji and Folk (1964), and by Harvey (1966).

We have undertaken an illustrative reconnaissance rather than a definitive analytical treatment, which could only result from more detailed work on specific rocks. A wide variety of petrographic and paleontologic information lies hidden in fine-grained limestones—clues to the origin of their constituent particles, to the age and environment of their deposition, to diagenetic cementation and recrystallization, and to the early stages of dynamic and thermal metamorphism. We have interpreted some of these clues, but others will no doubt see different information in these pictures. We hope above all that this little book will stimulate further work in ultramicroscopic petrology and paleontology.

[1] From *nanno-plankton*, the very small pelagic organisms which pass through the meshes of a plankton net, derived from *nanos*-dwarf (Greek).

ACKNOWLEDGMENTS

This investigation was made possible through a grant (1114-A2) from the Petroleum Research Fund administered by the American Chemical Society, which partly supported first Dr. Honjo and later Dr. Garrison as Postdoctoral Fellows, which supported Messrs. J. Murray, P. Temple, and D. Bukry as Petroleum Research Fund Fellows during part of their graduate studies at Princeton, and which covered running expenses and some capital equipment. Additional support for this work came from the Eugene Higgins Research Bequest to Princeton University. Grant 515-G2 from the Petroleum Research Fund enabled Garrison to continue work on West Coast Cenozoic and Mesozoic limestones. The Mesozoic limestones from the Alps and the Paleocene-Eocene limestones from the Spanish flysch were collected in the course of studies supported by the National Science Foundation (G 11855). It is a pleasure to acknowledge our debt to the donors and administrators of these Funds.

The specimens of limestone crusts from the Mediterranean sea floor, dredged by the Austrian Pola expedition, were kindly made available by Prof. H. Zapfe of the Naturhistorisches Museum, Vienna. Limestone crusts dredged off Barbados by the University of Miami's Institute of Marine Science were kindly provided by Drs. F. Kozcy and R. Hurley. Limestone samples of the Franciscan Formation were furnished by Dr. E. H. Bailey of the U.S. Geological Survey, Menlo Park, California, samples from the Devonian Waterways Formation by Dr. J. W. Murray of the University of British Columbia (Vancouver), and material from the Pacific Northwest by Dr. W. R. Danner of the same institution.

Our techniques were adapted or developed by Honjo. The actual photography was done by Honjo and later by Garrison, the printing by Honjo, Garrison, Mrs. Kazuko Honjo, Mrs. Jan Garrison, and Mr. G. W. Fischer. The instruments used were a Hitachi HS-6 and a Hitachi HU-11, both in Princeton University's Department of Biology. For the use of these instruments and the associated facilities we are deeply indebted to Dr. L. Rebhun. In addition to these persons and those listed above, we thank Dr. W. Gorthy of Dr. Rebhun's laboratory for help and advice and Mrs. H. Treumut for sample preparation. To Dr. T. Saito of the Lamont Observatory we owe special thanks for some foraminiferal dating of samples.

For the permission to include previously published figures, we are indebted to the following organizations:

To the American Association for the Advancement of Science for Figures 26, 54, 55, 57, and 60, out of Honjo and Fischer, 1964, *Science*, Vol. 144, No. 1620, pp. 837-839, copyright © AAAS 1964.

To W. H. Freeman & Co. and to the editors, for Figures 25, 26, 57, and 70, out of Honjo and Fischer, 1965, in *Handbook of Paleontological Techniques*, edited by Bernhard Kummel and David Raup, copyright © W. H. Freeman & Co., 1965.

To the Society of Economic Paleontologists and Mineralogists, for Fig. 64, out of Honjo, Fischer, and Garrison, 1965, *Journal of Sedimentary Petrology*, Vol. 35, pp. 480-88.

To the Alberta Society of Petroleum Geologists, the Edmonton Geological Society, and the Saskatchewan Geological Society, for Fig. 58, out of Garrison, 1967, *Bulletin of Canadian Petroleum Geology*, Vol. 15, pp. 21-49.

To the University of Chicago Press, for Figures 11-18, out of Fischer and Garrison, 1967, *Journal of Geology*, July issue.

CONTENTS

PREFACE   ix

ACKNOWLEDGMENTS   xi

LIST OF FIGURES   xiv

i · TECHNIQUES, IMAGES, AND
    INTERPRETATION   3
    PROCEDURE   3
    OTHER TECHNIQUES   3
    POSITIVE AND NEGATIVE IMAGES   6
    EXTRACTIONS, PSEUDOREPLICAS, ARTIFACTS,
        AND PROBLEMATIC STRUCTURES   8

ii · MINERAL GRAINS AND FABRICS   11
    CALCITE   11
    DOLOMITE   21
    PYRITE AND OTHER MINERALS   22

iii · FOSSILS   23
    "ALGAL" FILAMENTS
        OF UNCERTAIN AFFINITIES   26
    COCCOLITHOPHORIDA   26
    PROTOZOA: CALPIONELLIDS   37
    PROTOZOA:
        PLANKTONIC FORAMINIFERA   39
    MOLLUSCA: CEPHALOPODA   39
    TUNICATA   40

iv · PICTURES   41

REFERENCES   137

FIGURES

Frontispiece: Eocene of Olympic Peninsula, with discoaster

1 Schematic representation of replica method
2 Amoeboid and radiaxial fabrics
3 Recrystallization fabrics
4 Effects of thermal and dynamic stress on limestone fabrics
5 Fossils, and rim cementation
6 Coccolith limestone
7 Coccosphere of *Braarudosphaera bigelowi*
8 Sections of *Thoracosphaera* tests
9 Wall of *Calpionella*
10–16 Late Cenozoic lithified globigerinid ooze on present sea floor, off Barbados
17–18 Late Cenozoic lithified globigerinid ooze from Mediterranean sea floor
19–20 Eocene of Olympic Peninsula, Washington
21–22 Lower Eocene (Ypresian) flysch, Zumaya, Spain
23–36 Paleocene (Montian-Landenian) flysch, Zumaya, Spain
37 Cretaceous (Maestrichtian) chalk, Denmark
38–43 Cretaceous Franciscan Formation, California
44–45 Jurassic (Tithonian) Solnhofen limestone, Germany
46 Lower Cretaceous (Valanginian?) Rossfeld Beds, Austrian Alps
47–50 Lower Cretaceous (Berriasian) part of Oberalm Beds, Austrian Alps
51–64 Upper Jurassic (Tithonian) part of Oberalm Beds, Austrian Alps
65–66 Middle to Upper Jurassic Ruhpolding Radiolarite, Austrian Alps
67–70 Middle Jurassic (Bajocian) part of Adnet Beds, Austrian Alps

ELECTRON MICROGRAPHS
OF LIMESTONES
AND THEIR NANNOFOSSILS

Das Grosse und das Kleine, ausgemessen,
führt in die Welt der Sterne und Atome.
Nichts aber lehrt uns Fische tief im Strome
was unser Mass bedeutet zu ermessen

Anton Santer
Übung des täglichen Jenseit,
from *Verse und Reime,*
Wagner'sche Universitäts-Buchhandlung
Innsbruck, Austria, 1956

I· TECHNIQUES, IMAGES, AND INTERPRETATION

PROCEDURE

The pictures here published have been prepared by means of two-stage replicas (Bradley, 1961) of polished and etched surfaces. The procedure is described in detail in Honjo and Fischer (1965a) and consists of the following steps: (1) embedding of a suitable small block of limestone in a plug of epoxide resin; (2) grinding and polishing to produce a flat, smooth surface that cuts across the grains; (3) light etching of the surface, for about one minute in 0.1 to 0.05 normal HCl, mainly to form grooves along crystal boundaries and to bring out intracrystalline features such as twinning, cleavage, zoning, and inclusions; (4) replication of the etched surface on standard electron microscopy replicating tape; (5) chromium shadowing of this replica in a vacuum evaporator at an angle of about 30°, aiming at a chromium layer up to 20 Å thick; (6) backing this with a carbon layer about 200 Å thick, which, together with the metal, forms the second (positive) replica; (7) scoring the replica into squares of suitable size; (8) removal of the first (negative) replica in acetone; (9) standard procedures for collecting the little squares of the carbon-and-metal replica, placing on grids, and photographing. The manner in which the image is produced is shown diagrammatically in Fig. 1.

OTHER TECHNIQUES

The technique we used is but one of various techniques used by different investigators in this field, and the reader who wishes to make detailed comparisons of electron micrographs must bear the differences among the techniques in mind.

There is, first of all, the difference between fractured surfaces and surfaces that are polished and etched. Electron

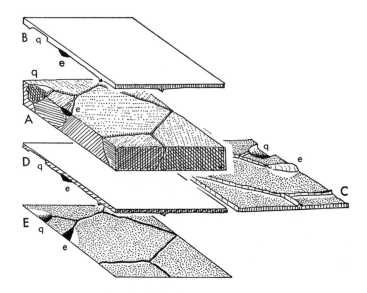

Fig. 1. Schematic representation of two-stage replication.

A. Limestone specimen, ground, polished, and lightly etched. Etching has grooved grain boundaries and has left non-carbonate grains (q and e) standing in relief.

B. First replica, an acetate film "peel." Grain boundaries form ridges, grain q is recorded as a depression in the replica; grain e has been bodily plucked out or "extracted."

C. First replica "shadowed" with metal, which highlights the "near" sides of raised features, such as grain e and grain boundaries, and the "far" sides of depressions such as grain q. Raised features throw uncoated "shadows." This is *negative relief*. The metal "shadow" is then backed with a uniform carbon film; the carbon and metal together form a second replica. First replica (peel) is now removed by solution.

D. Second replica in the electron microscope.

E. Image of second replica, as seen on fluorescent screen of electron microscope, or on prints made from film plates exposed to electron beam. Note that grain boundaries and the extracted grain e appear black and throw white shadows.

micrographs of fractured surfaces are exemplified by the work of Shoji and Folk (1964) and Flügel and Fenninger (1966) ; polished and etched surfaces are shown by Grunau and Studer (1956), Grunau (1959), and by our samples. We have tried both methods, and feel that the polished and etched surfaces yield more direct information on grain boundaries and on fossils, whereas the fractured surfaces bring out the grain surfaces and cleavage patterns and may provide a better idea of the grain size. Some of the surfaces prepared by us show a certain amount of fracture relief as a result of grain plucking in the polishing process.

One next has a choice between one-stage and two-stage replicas. In a pre-shadowed one-stage process, the fractured or polished-and-etched surface is shadowed and carbon-backed, then the rock is dissolved away and the replica collected, mounted, and photographed. This is the method used by Flügel and Fenninger (1966).

We have used the two-stage method, described in the previous section. Variations in the two-stage method involve primarily the choice of which stage to shadow. We have consistently shadowed the first stage, i.e., the negative surface of the replicating tape, prior to preparing the carbon film which forms the body of the second replica. Farinacci (1964) has used the same procedure. Shoji and Folk (1964), on the other hand, prepared a positive carbon film replica from the replicating tape and shadowed this second replica.

Each of these methods has its own advantages and disadvantages. We may look on these in terms of (1) faithfulness of the image, namely freedom from artifacts; (2) overall effect of the image; and (3) effect on details of relief.

The image is mainly conveyed by the shadowing. Inasmuch as each step in the procedure is likely to introduce some artifacts, the most faithful and trustworthy patterns are no doubt obtained in the one-stage method, in which the surface of the rock itself is shadowed. We did not use the one-stage method: there are technical difficulties in dissolving the rock and obtaining clean replicas; furthermore, the specimen is destroyed in the process, with no possibility of obtaining further replicas from it. Next in order of faithfulness is the shadowing of the replicating tape, the method we used. The method most liable to artifacts is the third one, in which shadowing is delayed to the very end (second replica).

POSITIVE AND NEGATIVE IMAGES

The replica technique is used mainly to convey a picture of the relief on a surface—whether the rough hills-and-valleys "landscape" of a fractured surface or the essentially flat but furrow-scored surface produced by polishing and etching. The same surface may yield very different looking micrographs, depending on the techniques of replication and photographic reproduction. The main variations are of two types: (1) whether the relief shown is positive or negative and (2) whether the tone in which it is reproduced is positive or negative. Yet a third variable, with which we shall not be further concerned here, is whether the image is a direct or a reversed (mirrored) one.

The electron beam cannot record relief directly; it can only depict differences in the electronic transparency of the replica. The relief is converted into such differences by the process of "shadow casting," which might have been more appropriately named "lighting." Metal, evaporated onto the surface from a point source at a definite angle, "highlights" the promontories and source-facing slopes with a thicker layer of metal than is deposited on lee slopes. Sharp projections cast definite shadows which are essentially free from metal deposits.

Positive and Negative Relief (Table 1)

The investigator can shadow the object itself—the "preshadowed single-stage replica process." In this way, he obtains *positive relief* (in the manner of Grunau and Studer, 1956). Projections from the rock surface cast shadows.

Table 1

RELIEF AND TONE

Relief	Original rock surface	Positive
	Shadowing of first replica	Negative[1]
	Shadowing of second replica	Positive[2]
Tone (if metal is considered equal to light)	Image on microscope screen	Negative
	Image in first photographic step (film)	Positive
	Image in second photographic step (print or 2nd film)	Negative
	Image in third photographic step (print from 2nd film).	Positive

[1] Two-stage replica method, first shadowed, then backed with carbon film.

[2] Two-stage replica method, first carbon-backed, then shadowed.

If he prepares a first replica and shadows this, he obtains *negative relief*; here the projections from the rock surface are recorded as grooves and pits and will not cast shadows, whereas the grooves and pits in the rock surface will appear as projections and will cast shadows. This is the process we used; it is illustrated in Fig. 1.

If he prepares a second replica from the first, he reproduces the positive relief of the original object; shadowing this second replica will provide an image of *positive relief* (in the manner of Shoji and Folk, 1964).

POSITIVE AND NEGATIVE TONE (TABLE 1)

If we consider the source of the metal in the shadowing process as analogous to a source of light, then the metallic portions of the replica record the highlights, and the metal-free portions the shadows.

When such a replica is seen in electron projection on the fluorescent screen of the electron microscope, the tone is negative: the metallic-rich parts form a dark image, and the

· 7 ·

"shadows" are bright. On a developed photographic plate, these relations are reversed: the metal-coated "highlights" are clear, the electron-transparent "shadows" are black, and the image is one of positive tone. Prints made directly from such photographic plates are again negative in tone; therefore, critical students of three-dimensional objects, such as free coccoliths, commonly prepare their final photographic prints not from the original photographic plate, but from a second film "negative" prepared therefrom.

Whether the tone of a given picture is positive or negative can be readily determined from the manner in which extracted grains or other pseudoreplicas are rendered. Since these inhibit passage of electrons, they appear in the manner of the metallic highlights: white in a positive image, dark in a negative one.

Total Image

In summary, such pictures as those by Grunau and Studer (1956), Shoji and Folk (1964), and Flügel and Fenninger (1966) show positive relief in negative tone. Promontories on the original surface cast shadows, but their "highlights" are black, and their shadows white. Extractions are black.

Our pictures and those of Farinacci (1964) show negative relief, in negative tone. This leads to a peculiar combination of features, illustrated in Fig. 1. Grooves in the initial surface appear as dark lines on the final print and therefore give the impression of grooves (an example of the familiar double negative equals positive) ; however, these grooves cast shadows, and these shadows are white. Projections from the initial surface cast no shadows; their highlights (in terms of metal source) are dark, their shadow side is bright.

EXTRACTIONS, PSEUDOREPLICAS, ARTIFACTS, AND PROBLEMATIC STRUCTURES

Mineral grains which adhere to the replica ("extractions") act as electron barriers. Carbonate extractions can be removed by acid and have been so removed in the work of Grunau and Studer (1956), Shoji and Folk (1964), and Flügel and Fenninger (1966); the extractions shown in their pictures are presumably mainly clay shreds. We have not normally treated our replicas with acid, and some of our extraction material may be carbonate, constituting "noise" rather than "information." However, our pictures of organic-rich limestones (Figs. 87, 88) show a clean contrast between the replicated carbonate grains and the extracted organic-rich matrix.

Spurious features due to remnants of the first (tape) replica (Fig. 84), or to doubling and shredding of the carbon film, may be a problem, especially when working on rough surfaces, but can generally be recognized as such.

Artifacts constitute the bane of electron microscopy; the investigator is commonly uncertain of whether a specific feature in a picture is related to a structure in the object or has resulted entirely from the replication and shadowing processes or from introduced particles. This is especially true when venturing out into a new field, such as limestones. Etch-resistant grains commonly retain scratches induced during grinding and polishing (Fig. 13). We suspect that the "marbled" patterns shown on calcite in some of our pictures (Figs. 21, 87, 88) are artifacts resulting from "spattering" of the carbon backing. We suspect that the "Spanish moss" pattern of Shoji and Folk (1964) may also be an artifact in the carbon replica; we have found nothing like it.

Another pattern which Shoji and Folk described on surfaces of calcite grains is one of tiny granules, commonly arranged in groups, which led them to the name "tea leaf pattern." These granules, as described and pictured by them, are always raised above the general level of their fracture surface and are not accompanied by corresponding pits.

Our polished-and-etched surfaces commonly show the same phenomenon (Figs. 21, 26, 27, 32, 36, 45): raised granules on the original surface. Due to our method, which yields negative relief by shadowing of the first replica, on which the granules are recorded as pits, the "tea leaf" structures look different in our prints. They are commonly but not inevitably present on grains of calcite spar or microspar, and may occur on coccolith plates, but we have not found them in rocks recrystallized under thermal stress. They are much more common in our earlier work than in our later pictures. "Tea leaves" may be directly or indirectly related to the structures of certain carbonate grains, but we are as puzzled as Shoji and Folk about their actual origin.

II· MINERAL GRAINS AND FABRICS

Our work has been done with replicas of etched surfaces, and the only mineral matter which has been exposed to the electron beam is that which accidentally adhered to the replica, in the form of extractions. We have therefore not been able to make direct specific determinations of the mineral grains shown on the photographs and have had to base our interpretations on indirect evidence, such as the bulk composition of the rock, the habit of grains, and their resistance to etching. Most of the rocks illustrated were also studied in thin section, and some were investigated with X-ray diffraction. For more detailed electron investigations of minerals it would be necessary to work with thin sections or powders a fraction of a micron in thickness, penetrable to the electron beam, or to combine electron microscopy of replicas with electron microprobe study of the original rock surface.

CALCITE

The rocks which we have investigated are presumably composed primarily of stable, low-magnesium calcite. Exceptions to this are seen mainly in the samples of Late Cenozoic pelagic limestones (Figs. 10-18), which are predominantly composed of high-magnesium calcite and contain some aragonite as well as low-magnesium calcite. Dolomite is present in some of these Late Cenozoic samples as well as in some of the older rocks studied, and is discussed in a succeeding section.

Calcitic grains may be divided into three types: (1) skeletal remains, (2) secondary chemical precipitates in voids ("cement"), and (3) a large number of grains of uncertain or heterogeneous origin. Assignment of any particular grain to one of these categories depends upon its size and shape, upon its relation to adjacent grains, and upon internal features of the grain; these parameters are revealed by etch patterns.

Etch Patterns

The pictures here presented are essentially pictures of etch patterns. The dominant element in these is the preferred etching along crystal grain boundaries—both external boundaries with adjacent grains and internal boundaries around inclusions. However, finer etch patterns develop within the individual crystal grains, and these are discussed first.

Many of the calcite grains show distinctive etch patterns, which appear to fall into four categories: (1) zoning or growth lines, (2) slip lines, (3) twinning lamellae, and (4) cleavage. Of these the growth lines are the most notable in undeformed or little deformed rocks.

Growth lines. Growth lines of a crystal which altered its habit repeatedly are shown in Fig. 63. Other examples of such growth bands are shown in the sparry grains filling coccospheres (Figs. 21, 27). A given grain will generally show two or more directions of growth banding, which will of course not cross.

Growth lines are well developed in many coccoliths. They are apparent on the surface-view electron micrographs published by other workers and appear in many of our pictures. In equatorial sections of ordinary coccoliths (Figs. 26, 53) they parallel the outer margin of each plate and thus form a concentric pattern as a whole. In cross sections of coccoliths they are not as often apparent, but in the centered section through a coccosphere (Figs. 5D, 21) they cut obliquely through the individual plates and maintain uniform directions through several adjacent plates. In *Braarudosphaera* they parallel the surface (Figs. 6, 31, 33).

Slip lines. Calcite grains in sheared rocks show sharp, straight furrows, which we interpret as slip lines along which the crystal lattice was displaced (Fig. 91).

Twinning. Twinning lamellae are less common or obvious than we expected; they appear in coarse and mechanically deformed rocks (Figs. 62, 63).

Cleavage. Cleavage is not as apparent in our cut preparations as it is in replicas of fracture surfaces; it is well shown in Fig. 22, in the large crystal filling a thoracosphere, and in the plates of the discoaster illustrated in the frontispiece.

SKELETAL GRAINS

The skeletal grains, which make up the larger part of some limestones but are absent from many other samples, are discussed in detail in a following section. They may occur in essentially unaltered form or in various stages of recrystallization, grading into indeterminate grains. Our work provides only a brief glance into the vast and as yet largely untravelled realm of ultramicroscopic structure in plant and animal skeletons. The most commonly encountered remains are those of coccolithophorids, nannoconids, calpionellids, and planktonic foraminifers, all thought to have been composed of low-magnesium calcite.

Sediments composed of very small skeletal materials of the above organisms are termed oozes, and in the fossil state, when not tightly consolidated, they are termed chalks (though many chalks are dominantly composed of calcite grains which are not recognizably organic, as in Fig. 37). Under pressure the individual grains are dissolved at their points of contact, coming to fit against each other along ever widening surfaces; the dissolved material may be deposited in adjacent voids. This process, by which the porous sediment is converted into a hard, brittle limestone of bulk density, is termed *solution welding.* It is illustrated in many of our pictures, especially in Figs. 6, 24-32, 48-60, and 65-67.

CHEMICAL VOID FILLINGS

Pore spaces in limestones are commonly filled by carbonate crystals growing inward from the walls. The fabrics of such fillings range from radiaxial (fibers growing commonly in sheaves or bundles), to mosaics of coarse grains, to single crystals. Radiaxial fillings are illustrated by the foraminiferal chambers in Figs. 2, 11, and 12, and show "Bathurst's rule" (1958)—an inward increase in grain diameter. Fibrous deposits in these examples are notably porous. In Fig. 17, a foraminiferal chamber is lined by much coarser crystals. Figures 29, 30, and 36 show foraminiferal and coccosphere chambers filled with more nearly equant grains. Figures 3C and 71-72 depict voids left by what we presume to be algal filaments in a dolomitic algal mat limestone; the voids were filled by coarse, irregular anhedral calcite crystals. Figures 21, 22, and 27 show cavities of coccospheres and calcispheres, each filled by a single calcite crystal.

Secondary enlargement of skeletal grains by calcite deposited in optical continuity ("syntaxis") with the calcite of the skeletal is termed *rim cement* (Bathurst, 1958). This is shown in many of the pictures, for example in Figs. 5A, 5B, 5C, 30, 31, 35, 36, and 52.

Recrystallized Grains and Grains of Indeterminate Origin

The shallow-water limestones are almost entirely composed of interlocked calcite grains, which retain no diagnostic organic structure. Similar grains make up parts of the organic oozes mentioned above. Presumably such fabrics include grains formed by the recrystallization of aragonitic muds and of unstable skeletal carbonates, as well as primary calcite grains and void-filling grains not recognized as such.

The Solnhofen lithographic limestones (Figs. 44, 45) are thought to have been aragonitic muds (Barthel, 1964) with some admixture of coccoliths (Flügel and Franz, 1967); the fabric shown is presumably the result of recrystallization to stable calcite.

The Late Cenozoic limestone crust from near Barbados (Figs. 10, 13-16) is a recrystallization product of a heterogeneous ooze which contained a mixture of low-magnesium and high-magnesium calcite as well as aragonite. Skeletons originally of low-magnesium calcite (Fig. 15) may show recrystallization features, and the bulk composition of the rock is now high-magnesium calcite (Fischer and Garrison, *in press*). Presumably the ultimate fate of this rock will involve a second recrystallization to stable, low-magnesium calcite. Thus the recrystallization fabrics shown by an ancient limestone may be the result of several successive episodes of recrystallization.

Among the most distinctive fabrics found is that of a highly bituminous and somewhat dolomitic limestone of a basin facies (Figs. 87-89). In this, rounded calcite grains and rhombs of dolomite are embedded in a matrix which appears to be largely composed of organic matter (Murray, 1966). The smooth etch on the calcite grains and the presence of overgrowth-like crystal faces on some grains indicate that these rounded grains are monocrystalline units, perhaps chemical precipitates (microspar) which developed anhedrally because of influence from the matrix.

INCLUSIONS

Inclusions of various kinds are widely shown throughout our photographs. Some are included foreign mineral grains such as pyrite (Fig. 64) and quartz (Figs. 62, 66, 84, 92). Iron oxides, silicates, and other minerals must be present in many of the sections, but we have not generally been able to determine them (Figs. 75, 79-82, 94). These grains are generally more etch-resistant than is the calcite and therefore stand out over the original calcite surface and are recorded as pits in the replica. Thus they do not cast shadows.

Another type of solid inclusion is produced by the engulfment of small calcite grains by larger ones of different orientation. This process is exemplified by the envelopment of small fossils, such as the tiny coccolith illustrated in Fig. 50.

However, the great majority of inclusions seen in our

pictures formed shadow-casting elevations on the replica, corresponding to pits in the polished and etched rock surface. Most of these are probably liquid inclusions—water bubbles in the calcite grains. While many of these are rounded in shape, some rocks (Fig. 90) are replete with "euhedral" or "subhedral" bubbles—negative crystals, such as those found in quartz.

The incidence of inclusions is related to a number of factors in the origin and history of the grains.

Sparry crystals grown in cavities appear to be essentially free of inclusions. However, the porosity in radiaxial fiber-spar fillings (Figs. 11-14) suggests that large grains formed by recrystallization of such fiber-spar may develop abundant inclusions. Skeletal calcite grains tend to have few inclusions. The greatest incidence of inclusions is found in calcite grains that appear to have originated from the diagenetic recrystallization of aragonite muds, such as the Solnhofen limestones, and possibly the Ordovician Marathon limestone.

Such primary or early diagenetic characters are strongly overprinted by later history. The undeformed, flat-lying limestones which have been investigated range from very rich in inclusions (sea-floor crusts, Figs. 13-16; Solnhofen limestone, Figs. 44-45) to intermediate (Devonian Waterways Formation, Figs. 85-89). Of the folded rocks, a few— the very mildly folded Cambrian of eastern Newfoundland (Fig. 92) and some of the Ordovician Marathon limestone of the Marathon belt in Texas (Fig. 90) —are rich in liquid inclusions, while the majority of folded limestones (the Spanish flysch, the North-Alpine Jurassic-Cretaceous and Triassic samples pictured here) are of intermediate character. The coccolith limestones of the Spanish flysch and of the Alpine Mesozoic show somewhat fewer inclusions than the originally aragonite-rich backreef limestones of the Alpine region. In more intensely stressed rocks, such as the mechanically strongly deformed specimen from the Marathon limestone (Fig. 91) and the thermally recrystallized rocks of the Olympic Eocene (Figs. 19-20), the Franciscan Cretaceous (Figs. 38-43), and the Triassic of the Vancouver area (Fig. 83), the liquid inclusions have been eliminated.

FABRICS

Grain sizes and fabrics are synoptically compared in Figs. 2 to 6. Sizes show a wide range. The finest grained rocks encountered are the dolomitic algal mat limestones (loferites), shown in Figs. 71 to 73, composed mainly of grains in the 0.7 to 1.5 micron range.

Grain shapes are also compared in Figs. 2 to 6 and show a marked variation. Rocks at one extreme, typified by the Solnhofen limestone (Figs. 2C, 44-45), are characterized by highly irregular, embayed, deeply interlocked grains, forming what we may term an *amoeboid mosaic*. At the other extreme, typified by the Franciscan Calera-type limestone (Figs. 4A, 4B, 42-43) and the Triassic limestone from British Columbia (Figs. 4D, 83), are rocks with simple, block-like grains, forming a *pavement mosaic*. Amoeboid mosaics appear to characterize relatively undeformed limestones, while pavement mosaics characterize rocks which have been subjected to thermal stress.

This difference in fabric may explain in part Fischer's observation, in the Alpine Triassic, that the unmetamorphosed limestones are generally very tough and ring under the hammer, whereas their metamorphosed equivalents—whether sheared or not—generally break more easily under the hammer, and do so with a dull sound.

The peculiar aptness of the Solnhofen limestones for lithographic printing is, of course, dependent in part on their purity, homogeneity, and comparatively fine grain. Our electron microscopy suggests that two other factors may also contribute materially to this aptness: one is the amoeboid nature of the grains, the other the large number of fluid inclusions. The amoeboid fabric shows a particularly great length of grain boundaries per unit area and, since these boundaries are grooved and pitted in etching, provides

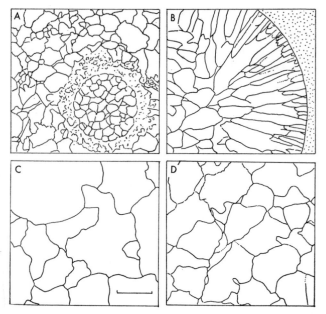

Fig. 2. Amoeboid and radiaxial fabrics, at equal magnification; scale bar on C 5 microns.

Chemical precipitation in cavities commonly leads to radiaxial fabrics with fibrous carbonate grains growing normally to the cavity wall. In A, a radiaxially filled *Thoracosphaera* has been tangentially cut, revealing cross sections of the fibers. In B, a radiaxially filled foraminiferal chamber has been cut parallel to the fibers, which conform to "Bathurst's rule" (1958, 1964) in growing larger from the wall toward the center. Amoeboid grain mosaics are shown by the matrix in A, and by C and D.

A. Limestone crust on the modern sea floor off Barbados of Late Cenozoic age (see Fig. 14).

B. Same (see Fig. 12).

C. Solnhofen lithographic limestone, Upper Jurassic, Solnhofen, Germany (see Fig. 44).

D. Fine-grained limestone in a neptunian dike, Upper Triassic Dachstein limestone, Tennengebirge, Northern Limestone Alps, Austria (see Fig. 75).

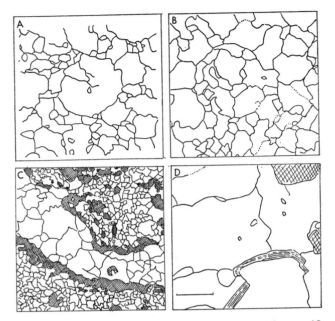

Fig. 3. Various recrystallization fabrics, at equal magnification; scale bar on D 5 microns.

A and B, coccolith (?) limestones with partial retention of fossil outlines. Mosaic remains comparatively ameoboid.

A. Red Lower Jurassic (Liassic) Adnet limestone from the Steinplatte, Austria.

B. Red Triassic (Carnian-Norian) Hallstatt limestones from the Feuerkogel, Austria (see Fig. 76).

C. Dolomitic algal mat limestone, with dolomite cross-hatched. Note the extremely fine-grained and anhedral nature of the calcite matrix (presumably recrystallized aragonite), and the similar nature of the dolomite. The more coarsely crystalline band running across the field is interpreted as a longitudinal section of an algal filament, which became preferentially coated with dolomite, became an open tube, and was then later filled with coarser calcite spar. Specimen from intertidal loferite facies of Upper Triassic Dachstein limestone, Steinernes Meer, Northern Limestone Alps, Austria (see Fig. 72).

D. Cambrian limestone from Avalon Peninsula, Newfoundland, showing coarse anhedral crystal mosaic, with clay minerals or mica flakes (lined) and quartz grains (cross hatched) (see Fig. 94).

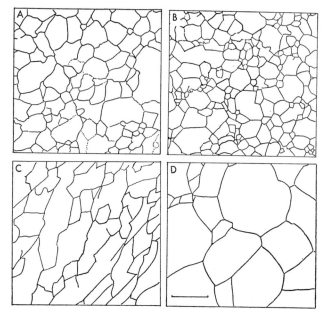

Fig. 4. Fabric effects of thermal and dynamic stress, shown at equal magnification; scale bar on D 5 microns.

A, B, and D are pavement mosaics, probably developed in response to thermal metamorphism. The loss of fluid inclusions is shown by the corresponding electron micrographs.

A. Comparatively uniform mosaic in Calera-type limestone from the Franciscan Formation, California.

B. Patchy recrystallization in what was probably a coccolith ooze, Calera-type limestone, Franciscan Formation, California.

C. Fabric showing strong preferred orientation of grain shape in Ordovician limestone from the Marathon fold belt, Texas. Note development of euhedral grain terminations in the direction of elongation (see Fig. 91).

D. Coarser pavement mosaic in recrystallized Upper Triassic limestone, Texada Island, British Columbia (see Fig. 83).

an abundance of pits, added to the very large number of liquid-inclusion pits already present. It is probably this high density of ultramicroscopic grooves and pits which provides the "tooth" that holds the ink in printing and gives the Solnhofen limestone its lithographic excellence.

Electron microscopy offers an approach to the study of preferred grain orientations in fine-grained rocks (d'Albissin, 1959). We have made no study of oriented sections, but call attention to the deformed fabrics shown in Figs. 40-41 and 91, the latter showing preferred grain orientation and slip lines, in an Ordovician limestone from the Marathon fold belt of Texas.

DOLOMITE

While we have not examined any pure dolomites, the mineral dolomite occurs as an accessory in many of the rocks studied. It normally appears in relief, due to its resistance to etching, commonly contains more inclusions than do the calcite grains, and occurs generally but not universally in euhedral rhombs. Such grains are shown in Figs. 18 and 87-89. They are presumably mainly grains developed in early or intermediate stages of diagenesis, formed from the magnesium supplied in part by the connate fluids, in part by the recrystallization of unstable grains of high-magnesium calcite. The rounded dolomite rhombohedron of Fig. 89 is noteworthy for its sharp growth zoning and its inclusions.

A different type of dolomite is illustrated in Figs. 3C and 71-72. This is a dolomitic algal mat limestone (loferite, Fischer, 1966), thought to be of an intertidal origin similar to that of the Persian Gulf (Illing, Wells, and Taylor, 1965) and the Bahamas (Shinn, Ginsburg, and Lloyd, 1965), where dolomite forms in very early diagenesis by replacement of aragonite. The dolomite in the Triassic loferites is exceedingly fine grained (average grain diameter is 0.7 microns), and the grains are anhedral.

While some of our Late Cenozoic deep-water samples

have shown dolomite in X-ray and electron micrographs (Fig. 18), others show rhombs (Fig. 13), but have yielded no dolomite peak on the X-ray (see also Fischer and Garrison, 1967). It seems possible that the rhombs shown here are a particularly calcium-rich protodolomite, or that they represent yet some other mineral.

PYRITE AND OTHER MINERALS

Among other minerals, the most distinctive is pyrite, which is brought out in relief by the polishing and the etching. Pyrite may occur in scattered cubes or other crystals and is probably responsible for the cuboidal black pseudoreplicas in Fig. 33. It also occurs in spheroidal aggregates termed framboids, shown in Fig. 64 (see also Honjo, Fischer, and Garrison, 1965).

No particular attempt has been made to identify quartz, clay minerals, etc., which presumably occur in many of our pictures. Possibly quartz grains are shown in Figs. 62, 65-66, 84. Clay minerals probably form some of the extraction pseudoreplicas in many of the pictures; they may compose some of the shredded matrix in Fig. 82, some of the "swirled" background in Figs. 87-88, and the intercrystalline flakes in Fig. 94.

Many of the limestones we studied are colored various shades of red; these are Eocene limestones from the Olympic Peninsula (Figs. 19-20); Paleocene limestones from Zumaya, Spain (Figs. 23-36); Laytonville-type limestones from the Franciscan Formation, California (Figs. 38-41); limestones from the Jurassic Ruhpolding Radiolarite of the Alps (Figs. 65-66); Jurassic Adnet limestones, Alps (Figs. 67-70); some Alpine Triassic shallow-water limestones (Figs. 71 and 75) and deep-water limestones (Figs. 76-78); and a Cambrian limestone from Newfoundland (Fig. 92). While red hematite grains of very small size are recognizable in thin section, we have not been able to identify them with certainty in the electron micrographs.

III · FOSSILS

Steinmann (1925) had noted that the hard, fine-grained Biancone and Majolica limestones (Jurassic-Neocomian) of the southern Alps and northern Apennines, when very thinly sectioned, showed multitudes of small round and elliptical bodies which he interpreted as coccoliths. He therefore viewed these very widespread limestones as fossil coccolith oozes, and suggested that many fine-grained and apparently unfossiliferous limestones of the geological column might in fact be fossil coccolith oozes, of comparatively deep-water origin. These coccoliths were, however, at the margin of visibility, and his documentation remained an unrealistic-appearing drawing.

About this time, *Nannoconus*, a probable coccolith of extraordinary size, was discovered in Jura-Cretaceous limestones of the Mediterranean-Alpine region, and occasionally micropaleontologists have recorded seeing some smaller coccoliths in the thin edges of thin sections, but the question of consolidated coccolith oozes was apparently not again taken up until Bramlette (1959), apparently not acquainted with Steinmann's work, reached a similar conclusion on different grounds. He pointed out that coccolith-rich chalks in Algeria change into hard limestones lacking recognizable coccoliths as they pass into the folded mountain belt—limestones most logically interpreted as consolidated and recrystallized coccolith chalks. By the same token, he argued that the North-Alpine Jurassic-Neocomian Aptychus Beds (our Oberalm Beds, equivalent in age and appearance to the Biancone and Majolica limestones discussed by Steinmann), correlative with coccolith-bearing marls in other areas and containing a pelagic fauna, probably represent a recrystallized coccolith ooze.

In our own optical work, we arrived at a position between Steinmann's and Bramlette's: Our thinnest sections of the Paleocene limestone at Zumaya, Spain (Fig. 23), clearly show coccospheres (*Thoracosphaera*) and giant coccoliths (*Braarudosphaera*), and also occasional specimens of normal coccoliths (but without sufficient detail to make closer identifications); the great bulk of the rock, however, remains a vermicular mass of small grains, unidentifiable even when sections are cut to a thinness of 2-5 microns. In the Alpine Oberalm limestone (Fig. 47) we could distinguish no coccoliths.

Electron microscopy, on the other hand, shows that both of these rocks (Figs. 24-36, 47-64) teem with heterogeneously oriented, embayed and solution-welded coccoliths, which, in the case of the Oberalm Beds, are partly recrystallized yet still distinctive.

Fine-grained limestones do not lend themselves to good optical resolution because (1) they are composed of interlocked grains of similar composition, lacking thus inherent contrast; (2) they are optically highly complex, the light being doubly refracted in each crystal, hence repeatedly refracted and reflected as it passes through adjoining and overlapping grains (we can recognize the individual grains by their optical character in cross-polarized light, but their boundaries are then diffuse); (3) although the component grains may once have had distinctive shapes, these have become so modified by solution welding that they are rarely diagnostic.

The replica method avoids the first two difficulties: the acid brings grain boundaries into sharp relief, which can be captured on acetate film and emphasized by metal shadowing. Such replicas yield crisper optical images than do thin sections, but the proper techniques for their study at high magnifications are yet to be developed.

The electron microscope combines the advantage of the replica with a resolving power going far beyond that of optical systems. The main identifying feature of solution-welded coccoliths, for example, lies in their internal structure of minute, complexly overlapped calcite plates, which lie beyond the resolution of optical microscopes.

Paleozoic limestones examined by us included samples from the fine-grained, red Lower Cambrian limestones of the Avalon Peninsula, Newfoundland; from graptolite-bearing Middle Ordovician limestones from the Marathon Mountains, Texas; from the Ordovician Salona limestone, Pennsylvania; from Silurian graptolite-bearing limestones from Bathurst Island in the Canadian Arctic; from the peri-reef and reef facies of the Devonian in Alberta; from the Pennsylvanian of the Mid-Continent region; and from the Permian Bone Spring Formation of the Delaware Basin, Texas. The only fossils revealed in this suite of samples are a few sections of megafossils, such as those in Figs. 84 and 86. Presumably many of the carbonate grains seen in the other samples are of organic origin, but were not recognized as such by us. Clearly this is, in part, a matter of insufficient sampling. Small fossils such as hystrichospheres, Chitinozoa, calcispheres, etc. are abundant in many Paleozoic rocks, and would be recognizable in low-power electron microscopy. However, our failure to find coccoliths or other calcareous nannoplanktonic bodies in the Paleozoic limestones, which matches the failure of coccolith workers to find them in Paleozoic shales and marls, tends to confirm the suspicion that the Paleozoic largely lacked a calcareous nanno-plankton.

The earliest suggestions of such nannoplankton in the rocks we examined are found in Triassic limestones of the Alpine Hallstatt facies (Figs. 76-77, 79-82). Here, some rocks show coccolith-like outlines, while others show complexly built grains which may or may not be fossils.

From the Jurassic onward, coccoliths are abundant, especially in deeper-water limestones. This was known from work on calcareous shales, but we have found fine-grained, hard limestones in the upper Adnet Beds of the Alpine Middle Jurassic, in the Oberalm Beds of the Alpine Upper Jurassic, in the Cretaceous of the Franciscan Formation, California, in the Paleocene and Eocene of the Spanish flysch, and in the Eocene of the Olympic Peninsula, Washington, to be essentially coccolith oozes.

Our sampling of shallow-water limestones was more restricted, but yielded fossil algal filaments in the Alpine Triassic and rare coccoliths in the Jurassic Solnhofen limestone. Flügel and Franz (1967) found coccoliths in greater abundance in the Solnhofen limestone.

The following pages provide data for the various groups of fossils encountered, listed in phyletic order.

"ALGAL" FILAMENTS OF UNCERTAIN AFFINITIES

Figures 3C and 71-72 illustrate calcite-filled tubes in intertidal, dolomitic, "algal mat" limestone (loferite) of the Alpine Upper Triassic. We consider these to be fossil "algal" filaments, belonging presumably either to the Cyanophyta or to the Chlorophyta (Fischer, 1966).

COCCOLITHOPHORIDA

The extremely small, biflagellate chrysomonadine algae termed Coccolithophorida are among the most abundant organisms in the present oceans, and their complex skeletal plates, loosely termed coccoliths and rhabdoliths, compose a large part of the present calcareous deep-water oozes. These skeletal remains are by far the most abundant fossils encountered in our survey of fine-grained limestones.

"NORMAL" COCCOLITHOPHORE REMAINS (HELIOLITHS)

The earliest generally accepted coccoliths are from Lower Jurassic rocks. Our earliest record of them, in this survey of limestones, is from lower Middle Jurassic (Bajocian) limestones at the top of the Adnet Beds, in the Unken Valley of the Alps (Figs. 67-68). Here they occur in rock-forming numbers, in a red, manganiferous limestone sequence which immediately underlies a "radiolarite" section. The "radiolarite," termed the Ruhpolding Radiolarite in

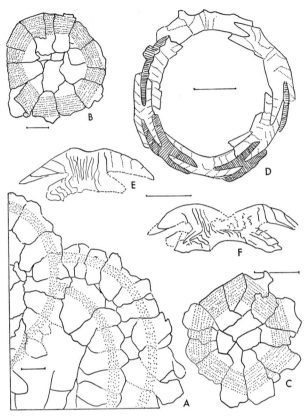

Fig. 5. Fossils and rim cementation; scale bars 5 microns.

A. Part of a globigerinid foraminifer (wall stippled), showing rim cementation on inner and outer surfaces of walls
and the development of a blocky crystal mosaic, controlled
by initial radial orientation of calcite in the shell wall.
Paleocene flysch, Zumaya, Spain (see Fig. 35).

B and C. Cross sections of whole coccospheres of *Braarudosphaera bigelowi*. Each of these sections cuts six individual coccoliths (specifically, pentaliths), which are shown by
stippling parallel to their growth lines. Each such pentalith
is in turn composed of five individual segments, of which
two to three have been intersected in each case. Each such
segment is a separate crystal unit and may show a rim-
cement extension toward the outside or toward the inside.

this region of the Alps, is in part radiolarian chert, in part red, hard siliceous limestone with abundant radiolarian tests, and is of Middle to Late Jurassic age; the limestones are extensively recrystallized but contain a few residual fragments of coccoliths (Figs. 65-66).

Coccoliths and other nannoplanktonic fossils again become extremely abundant in limestones of the Upper Jurassic–Lower Cretaceous (Tithonian–Berriasian) Oberalm Beds (Figs. 48-59), which overlie the Ruhpolding Radiolarite in the Unken Valley (Honjo and Fischer, 1964; Garrison, 1964, and *in press*).

Mid-Cretaceous rocks of North America also contain coccolith limestones. These are fine-grained, red limestones of the Laytonville type (Figs. 38-41), which are associated with pillow lavas in the Franciscan Formation of California (Garrison and Bailey, *in press*).

Fine-grained limestones are a conspicuous part of the Upper Cretaceous to Lower Tertiary flysch of Europe—for example, in the flysch of the North-Alpine rim in Austria, in the Monte Antola flysch of the northern Apennines, and in the Pyreneean flysch at Zumaya, Spain. The latter flysch sequence was investigated by us, and Figs. 5A-C, 5E-F, 6, and 23-36 show pictures of a Paleocene red limestone from this sequence. Figures 5D and 21 show a complete coccosphere in a grey Eocene sample from Zumaya. These flysch

---

Paleocene flysch, Zumaya, Spain. See also Figs. 6, 31, and 33.

D. Another type of coccosphere (*Coccolithus* sp.), from the Eocene at Zumaya, Spain, showing the interlocking of true coccoliths. Alternate coccoliths shaded to clarify structure. Plates within individual coccoliths outlined by stippling (see Fig. 21).

E and F. Individual coccoliths (*Coccolithus* sp.) in vertical cross section. Note individual plates with coccoliths. Paleocene, Zumaya, Spain (see Fig. 27).

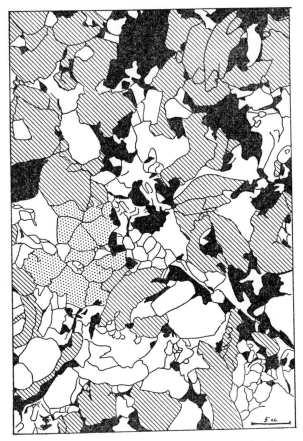

Fig. 6. Solution-welded fabric of coccolith limestone, Paleocene, Zumaya, Spain. Ordinary coccoliths insofar as recognizable by plate structure are hachured, *Braarudo-sphaera* remains stippled. Unidentified carbonate grains shown in white, extracted grains (clay, carbonate) in black.

limestones are mainly coccolith oozes, representing either un-disturbed particle-by-particle "background" material de-posited between turbidite influx, or such sediment stirred up and redeposited as parts of turbidites.

We have made no attempt to identify the majority of coccoliths with described genera or species. This is due partly to our lack of experience in coccolith taxonomy, and partly to the circumstance that our pictures are mainly sections through coccoliths rather than surface views, and are not directly comparable to the latter. One of the difficulties with sections is that one cannot be sure whether one sees a coccolith "right-side up" or "upside-down," and one cannot, therefore, determine whether it spirals to the right or left. In mixed assemblages one may not know which vertical sections correspond, in species, to which equatorial sections. On the other hand, the sections provide information on the internal structure of coccoliths.

Family Braarudosphaeridae

Some coccolithophore remains are so distinctive that they may be generically and even specifically determined in our sections. The most distinctive of all are the Braarudosphaeridae Deflandre, characterized by rather pyritohedron-shaped coccospheres (Fig. 7, reconstruction). Each such coccosphere is composed of twelve flat-faced, pentagonal coccoliths (pentaliths), each of which is, in turn, composed of five radial, monocrystalline segments.

The genus *Braarudosphaera* Deflandre is known to range from Early Cretaceous to Recent. All of our illustrated material belongs to the well-known *B. bigelowi* (Gran and Braarud) and was obtained from the sample of the Spanish Paleocene flysch at Zumaya. Pentaliths range in diameter from 5 to 20 microns. Odd sections through pentaliths or fragments of pentaliths are shown in nearly every picture we have made of this sample. An optical view is provided in Fig. 23, an equatorial section of a pentalith in Fig. 24, tangential sections of coccospheres in Figs. 32-33, and cross sections of coccospheres in Figs. 5 and 31. Both pentaliths and coccospheres are large enough to be readily apparent in very thinly cut thin sections. *B. bigelowi* is distinguished from other *Braarudosphaera* species by having its pentaliths

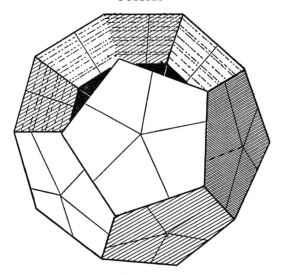

Fig. 7. Coccosphere of *Braarudosphaera bigelowi*, with two of the pentaliths removed so as to show central cavity, and the laminated edges of three pentaliths. Width of pentaliths ranges from 5 to 20 microns in this species. (Modified from Deflandre.)

composed of five equal radial plates, the sutures of which, radiating out from the center of the pentalith face, reach the margin neither at the corners of the pentagon, nor at the midpoints of the sides, but in an excentric position.

Another characteristic of the pentaliths, well illustrated by our material, is the lamination parallel to the outer or inner surfaces, also mentioned by Bramlette and Martini (1964) and by Hay and Towe (1962). Contrary to implications by Bramlette and Martini, the lamination does not disrupt the crystalline unity of the individual pentalith segments, but is an intracrystalline effect. We assume that this lamination results from growth, but do not know whether growth proceeded toward the inside, toward the outside, or both.

FAMILY DISCOASTERIDAE

The family Discoasteridae Tan Sin Hok comprises star-shaped nannofossils in which each ray is a separate crystal unit. The stars are in the size range of coccoliths, averaging 10 microns in diameter, but their simple crystal structure makes them more accessible to study by optical microscopy. Discoasters have been rarely reported from the Cretaceous, are abundant in Tertiary sediments, and became nearly extinct round about the beginning of Pleistocene time. We have encountered them in the Eocene of the Olympic Peninsula: The curious radial aggregates of Fig. 19 are probably oblique sections through discoasters, and the frontispiece shows an equatorial section through the common Eocene species *Discoaster barbadiensis*. Note on this the prominent cleavage patterns, which show each ray to be a crystal unit.

FAMILY THORACOSPHAERIDAE

The nannoplankton of today contains calcareous spheres composed of pavement-block-like grains of calcite fitted together in a tightly interlocked mosaic. These grains, and the wall which they form, are from 1 to 2 microns thick, and the diameter of the sphere lies between 10 and 20 microns. The sphere may be complete, or may have one open pole, presumably for the exsertion of flagellae. The individual blocks of the wall are each pierced by an axial pore. Kamptner (1927) interpreted these spheres as coccolithophore remains (coccospheres) and erected the genus *Thoracosphaera* and the family Thoracosphaeridae. This interpretation has been upheld by Deflandre (1952) and various other coccolith workers, but cannot be considered as proved inasmuch as the live organism has not yet been described.

Coccolithophores are not the only organisms which produce small, hollow calcareous spherules. Dasycladacean algae such as *Acetabularia*, for example, produce calcareous spore capsules (Rupp, 1966), and small calcareous spherules in the fossil record have been described as Foraminifera.

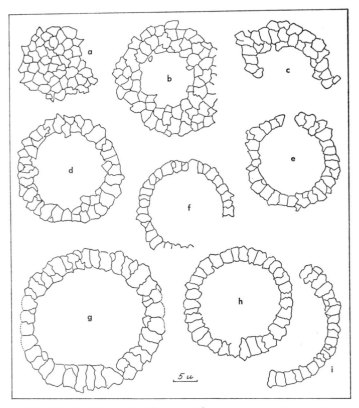

Fig. 8. Sections of *Thoracosphaera saxea*, drawn from electron micrographs. Paleocene, Zumaya, Spain. *a* is a strictly tangential section; *b, c,* and *d* are sections progressively closer to the center of the coccosphere, and *e-i* are essentially centered sections. Compare Figs. 27, 28.

Most such fossil spherules described to date have been seen in thin section and show diameters of 60 to 200 microns and walls 10 to 30 microns thick, about an order of magnitude larger than the Recent *Thoracosphaera*. Most have been referred to the problematic genus *Calcisphaera* Williamson (Baxter, 1960), the wall structure of which has not been described in detail. Such "calcispheres" have been found in every system from the Ordovician onward and are probably not of coccolithophorid origin.

Our electron microscopy has revealed, in various rocks, spheres with diameters of 10 to 40 microns, with wall thicknesses of 1.5 to 3 microns, and with the blocky wall structure characteristic of *Thoracosphaera*. The earliest possible structures of this sort are in Triassic Alpine Hallstatt rocks (Fig. 67) but are too poorly preserved to permit more definite treatment; somewhat better ones occur in the Mid-Jurassic (Bajocian) Upper Adnet limestone (Figs. 69-70). Well-preserved thoracospheres occur in the Spanish Paleocene and Eocene, and in Late Cenozoic limestone crusts on the sea floor off Barbados.

Like braarudospheres, thoracospheres are large enough to be readily seen in thin sections. In very thinly cut sections, the comparatively coarse structure of their walls is apparent, although the number of component elements can generally not be determined. The shape and structure of the individual calcite blocks is not revealed by optical means in thin section, but is visible in free specimens.

Thoracosphaera saxea Stradner, 1961 (*T.* cf. *imperforata* Bramlette and Martini, 1964). The Paleocene at Zumaya contains abundant thoracospheres characterized by diameters of 20 to 30 microns; a wall thickness of about 3 microns; and rather coarse wall blocks, from 2 to 4 microns broad, and imperforate (Figs. 8, 24). Tangential sections (Figs. 8, 25) bear out the imperforate nature of the blocks and show that in plan they are irregularly shaped and sutured together in a somewhat complex manner. A normal cross section intersects 30 to 40 blocks.

Stradner's drawing of the type, from the Danian of Austria, seen only optically, shows even more elongate blocks and more intricate zig-zag sutures. Bramlette and Martini, who record their *T.* cf. *imperforata* from the type Danian and from the Maestrichtian Arkadelphia clay of Arkansas, illustrate a form much more like ours; however, they mention that topotype material from Stradner's *T. saxea* locality is less intricately sutured than his picture. We are proceeding on the assumption that the extreme elongation of blocks shown by Stradner, as against the more equant structure found by Bramlette and Martini and by our electron microscopy, is more apparent than real. If pairs of blocks were to have nearly but not perfectly equal crystallographic orientation, then they would appear as a single long block in polarized light, and might appear linked in normal light, but would appear as separate units in replicas made from etched surfaces.

Thoracosphaera cf. *T. heimi* (Schiller). The same red limestone at Zumaya also contains a rare *Thoracosphaera* (Fig. 30) which has a large, thin-walled shell composed of small blocks showing the axial perforation that characterizes the Recent *T. heimi*. The sphere is larger than any reported for the latter species, but a closer comparison cannot be made in view of our lack of a tangential section of our form and electron micrographs of *T. heimi*.

Thoracosphaera (?) sp. Figure 29 shows a circular section, in the size range of thoracospheres, which has a thick wall composed of prismatic elements. This specimen is from the red Paleocene limestone at Zumaya, but a thin section of the grey Eocene limestone at Zumaya also shows what appears to be a thoracosphere with a thick, prismatic wall.

Thoracosphaera cf. *deflandrei* Kamptner. The Eocene at Zumaya has provided us with a single electron micrograph (Fig. 22) of a large *Thoracosphaera* with many small wall blocks (about 70 shown in this presumably centered cross section). The blocks are imperforate. In the absence of a tangential section we remain ignorant as to their shape in plan, but the large number of elements and their imperforate nature resemble Stradner's (1961) picture of the above-named species.

Thoracosphaera sp. The lithified Late Cenozoic globigerinid ooze dredged off Barbados (Fig. 14) contains small and comparatively thick-walled calcareous spheres which fall into the size range of Recent thoracospheres, but which do not show the characteristic blocky structure of the wall. In optically examined peels (Fischer and Garrison, 1967, plate 4B) similar spheres show the typical *Thoracosphera* wall, and we therefore assume that the specimens studied under the electron microscope had undergone a retrogressive recrystallization of the wall, as have some of the globigerinoids in this rock. The species is probably referable to *T. heimi* or *T. imperforata*, but we cannot tell whether it was perforate or not.

FAMILY NANNOCONIDAE

The problematical genus *Nannoconus* Kamptner is large enough to be readily studied in thin section, and was first described on the basis of optical studies. Several species have been described. It is widespread in rocks of latest Jurassic (Tithonian) and Early Cretaceous age. Grunau and Studer (1956) published the first electron micrographs, and Farinacci (1964) has published others. It consists of a cone of calcite prisms radiating out from a central canal. The common Tithonian to Neocomian *Nannoconus steinmanni* is illustrated in Figs. 42, 49, and 50. A possible relative, differing in that it has a smaller number of prisms and in their irregular inner terminations, is illustrated in Fig. 49.

DOUBTFUL COCCOLITHOPHORE REMAINS

The Triassic limestones studied have yielded some structures which are nannofossils but are too badly recrystallized to permit any more definite assignment. Such are the Carnian and Norian fossils depicted in Figs. 76-77.

The essentially unrecrystallized Norian Pötschen limestone at Pötschen Pass has yielded pictures (Figs. 79-81) of ultramicroscopic bodies which are composed of batteries of small plates. We are uncertain at this time whether these represent grains of some platy silicate mineral, or whether they represent nannofossils. They are distinct from anything we have seen in specimens or in the literature.

PROTOZOA: CALPIONELLIDS

Calpionellids are small, vase-shaped tests normally studied under the optical microscope; they are found mainly in Tithonian and Neocomian limestones, although they are also known from the Devonian and Carboniferous. In their size and form, they resemble the lorica of tintinnid ciliates. However, the normal tintinnid lorica is agglutinated out of heterogeneous small silt grains, whereas the test of calpionellids is calcitic. Some workers have suggested that it is built of small carbonate grains picked up by the organism. Noel (1958) illustrated calpionellids which are plastered with coccoliths and suggested that they fed on coccolithophores where these were abundant and incorporated the coccoliths in the test. On the other hand, the calpionellid test normally shows a very fine fibrous structure with extinction normal and parallel to the wall (Bonet, 1956; Doben, 1963; confirmed in our sections). Electron microscopy (Figs. 9, 52) shows that the test is preserved much like that of the pelagic Foraminifera in the Zumaya Paleocene: The wall is now composed of radially oriented grains which are almost indistinguishable in the center of the wall, but show sharper boundaries toward the surface. The inner and outer surfaces are quite irregular, as one might expect as a result of syntaxial rim cementation and solution-welding with adjacent grains. A distinctive feature of the calpionellid test, which has its parallel in the globigerinid Foraminifera, is the appearance of a central band, in the middle of the shell, which is raised relative to the adjacent surfaces. This central band results from differential etching, and implies the existence of a central lamina of a composition distinct from the remainder of the test—perhaps richer

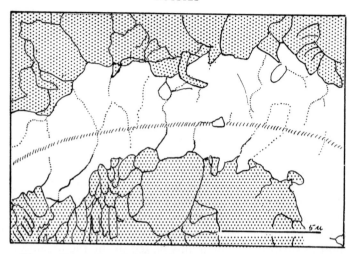

Fig. 9. Part of a *Calpionella* wall (Fig. 52). Wall white, matrix stippled. The originally fibrous wall has become somewhat reorganized into faintly defined, radially oriented crystals, which interlock with the grains of the matrix. The hachured band represents an etch-resistant zone in the middle of the wall, similar to an equivalent structure found in the globigerinid Foraminifera. In upper middle of picture, a foreign grain appearing to be part of a very small coccolith extends hook-like into the wall, possibly in illustration of Noel's contention that *Calpionella* agglutinated coccoliths to the surface of the test. Upper Jurassic Oberalm Beds, Unken Valley, Austria.

in organic matter. There is, then, no doubt that the calpionellid shell as such was chemically secreted by the organism. If foreign particles were incorporated into the test, they were only added to the surface, in the manner in which the snail *Xenophora* builds bits of other shell matter into its skeleton. The deep extension of a coccolith into the shell, shown in Fig. 52, suggests that this did indeed occur.

PROTOZOA: PLANKTONIC FORAMINIFERA

Planktonic Foraminifera from Paleocene rocks are shown in Figs. 5A, 23, 30, 35, and 36, and from Upper Tertiary or Quaternary limestone crusts of the modern sea floor in Figs. 11-12, 13, 15, and 17. These tests show a structure much like that of calpionellid tests, described above. Etching commonly leaves a ridge in the middle of the wall (Figs. 35, 36). Also, there is a tendency for the originally extremely finely fibrous walls to recrystallize into large, blocky grains of calcite (Figs. 5A, 35). As in the case of the calpionellid test, such grains are only faintly defined in the mid-parts of the wall, but their boundaries become more apparent toward the wall margins, and many such grains extend beyond the wall margin, in syntaxial rim overgrowths.

Figures 13 and 15 show another type of recrystallization: here the test is becoming reorganized into small grains which blend with similar grains of the matrix. This is a special case of lithification on the sea floor, with development of high-magnesium calcite (Fischer and Garrison, 1967).

MOLLUSCA: CEPHALOPODA

In the Permian Bone Spring limestone of West Texas, we interpret a thin, finely laminated shell wall to be that of a small cephalopod (Fig. 84), possibly retaining its original skeletal composition of aragonite. The "brick-and-mortar" structure is very similar to that of aragonite plates in organic matrix, in the nacreous layer of mollusks (Grégoire 1957, 1959).

A similar cross section through a somewhat recrystallized shell, probably also of a cephalopod, is seen in Fig. 78, from a red, Upper Triassic limestone of the Alpine Hallstatt facies.

TUNICATA

The Upper Cenozoic limestone crust from the sea floor near Barbados yielded a striking star-like cross section of what we take to be a burr-shaped, calcitic tunicate spicule (Fig. 16). The individual rays of the burr are composed of longitudinal fibers and appear to be quite porous.

IV· PICTURES

The electron micrographs and the accompanying descriptive material have been arranged in stratigraphic order, starting with near-Recent sediments and ending with the Cambrian. The original plates were mostly made at the lower limits of the microscope, at 1,000 to 1,500×, and were enlarged to various magnifications, mostly to 3,000 and 5,000×. The approximate magnification is shown on each picture by a white 5-micron scale bar. The observer should bear in mind that he is looking at negative relief, shown in negative tone (see Chapter I). Most of the magnifications are correct to within 10 per cent; however some pictures may be further out of scale as a result of using several instruments and several operators and using the Hitachi HU-11 both with and without its intermediate lens.

***Figs. 10-16. Upper Cenozoic (Miocene or younger) lithi-
fied globigerinid ooze on present sea floor, off Barbados.***

Lat. 12°52′N, Long. 59°32′W, depth 280-440 meters.
Dredged by University of Miami CARIBARC Cruise II.

Hard, fine-grained, conchoidally fracturing limestone
containing nodules and pancakes of manganese and iron
oxides. Probably a surficial crust on the sea floor. The fossil
assemblage is dominated by pteropods and other presumably planktonic gastropods, and by planktonic foraminifers,
but also contains coccoliths and fragments of echinoderms
and coralline algae. The mineral assemblage is dominated
by high-magnesium calcite (composition 12.5 mol %
$MgCo_3$), but also contains small quantities of low-magnesium calcite (globigerinid tests) and aragonite (portions of
gastropod shells which have not been altered to calcite).

Optical and electron microscopy show that this sediment,
originally a globigerinid ooze, has been altered by processes
which include partial removal of aragonite, filling of cavities
with radiaxial calcite, and a recrystallization of the matrix
and many of the calcitic fossils to a fine mosaic of high-
magnesium calcite. Reference: Fischer and Garrison, 1967.

Fig. 10. Optical photomicrograph of a peel replica, 100×,
showing globigerinid Foraminifera in varying stages of
growth and preservation, in fine-grained carbonate matrix.
Clear angular objects are volcanic shards. Photo Garrison.

Fig. 11. Inner chambers of a globigerinid test, showing
chamber fillings of radiaxial calcite. Due to tangential orientation of the cut, the fibers are transversely intersected (cf.
Fig. 12). Photo Garrison. From Fischer and Garrison, 1967.

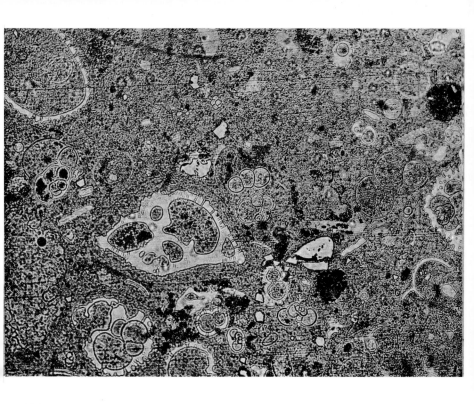

SCALE BAR 5 MICRONS

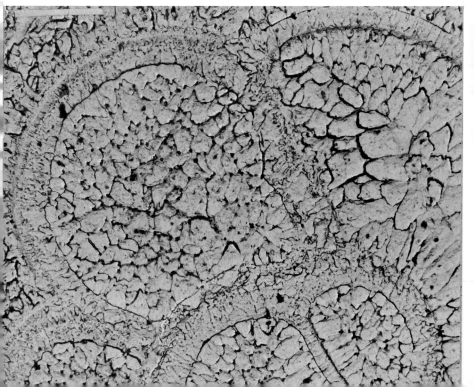

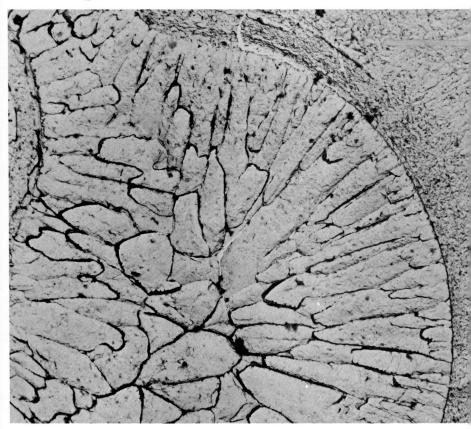

Fig. 12. A globigerinid chamber filled with radiaxial calcite showing "Bathurst's rule" (1958, 1964) —an increase in grain size from the margin toward the middle. Cf. Figs. 2B and 11. Photo Garrison. From Fischer and Garrison, 1967.

Fig. 13. Unidentified authigenic mineral grain (dolomite?) projecting on right into wall of a globigerinid foraminifer, which shows granular recrystallization. At upper left, cross section of a slightly curved coccolith. Photo Garrison. From Fischer and Garrison, 1967.

Fig. 14. *Thoracosphaera* with interior filled by radiaxial calcite, the fibers of which are transversely intersected. Note recrystallized matrix. Cf. Fig. 2A. Photo Garrison. From Fischer and Garrison, 1967.

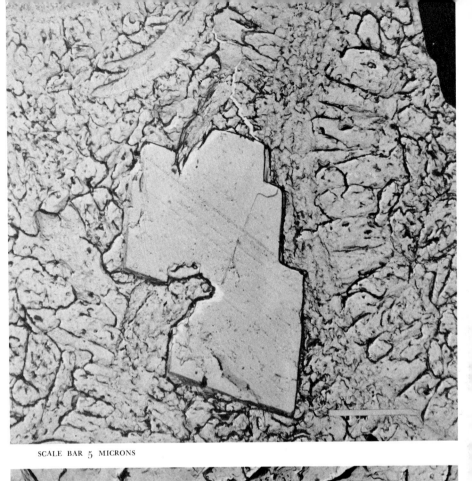

SCALE BAR 5 MICRONS

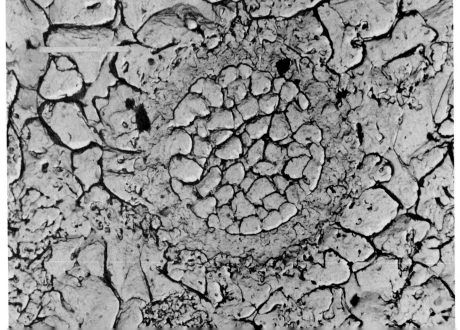

Fig. 15. Recrystallization attacking the walls of a globigerinid foraminifer. The chamber on the right shows a sharply defined boundary between wall and chamber filling. The chamber at the upper left shows a similar relationship on its right side, but on its lower side the boundary becomes blurred toward the left and individual calcite grains extend across it. In the chamber at the lower left, the boundary between test wall and matrix has become obliterated by granular recrystallization which involved both test and matrix. Photo Garrison. From Fischer and Garrison, 1967.

Fig. 16. A star-shaped tunicate spicule showing fibrous and porous structure of rays. The darker, central area probably represents oblique cross sections of other rays in a slightly excentric section. Photo Garrison. From Fischer and Garrison, 1967.

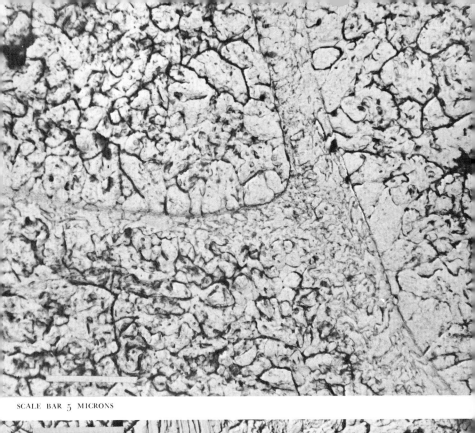

SCALE BAR 5 MICRONS

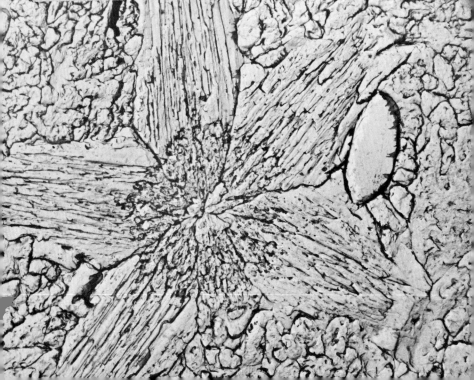

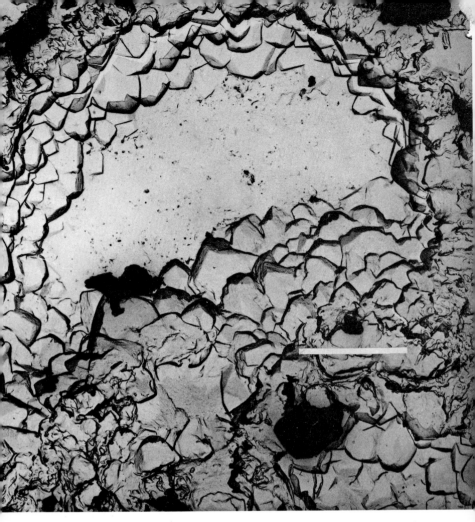

Fig. 17. Chamber of a globigerinid foraminifer, with drusy growth of calcite into largely hollow interior of chamber. Photo Garrison.

Figs. 17-18. Upper Miocene globigerinid limestone from Mediterranean sea floor.

Lat. 31°38'37"N, Long. 28°51'19"E, depth 2,055 meters, Austrian Pola Expedition No. 2, sample no. 85.

Surficially indurated crusts of globigerinid ooze, 1 cm. thick; upper surface stained with iron and manganese oxides. Mineral assemblage dominated by high-magnesium

Fig. 18. Dolomite rhombs in an inhomogeneous matrix of calcite grains. Photo Garrison.

SCALE BAR 5 MICRONS

calcite (around 7 mol % $MgCo_3$), with minor amounts of low-magnesium calcite and dolomite. This crust is far less thoroughly indurated than is the Barbados limestone described above, and the filling of foraminiferal chambers has not gone to completion. But all aragonite has been removed and dolomite has been formed in detectable quantities. From Fischer and Garrison, 1967. Age determination by T. Saito (pers. corres.).

· 49 ·

Fig. 19. Radial aggregates of calcite grains suggest organic origin, but are unlike anything we have encountered in other rocks. They are possibly discoasters or a *Braarudosphaera*-like coccolith. Matrix of finer carbonate grains and clay minerals. Photo Garrison.

SCALE BAR 5 MICRONS

Fig. 20. Coccolith remains and some non-carbonate grains.
Photo Garrison.

These fine-grained, hard, manganiferous limestones are usually red and contain abundant tests of planktonic foraminifers. They are most often associated in the field with pillow lavas and other products of submarine volcanism. The specimen here illustrated is from the Mt. Angeles area in the northern Olympic Peninsula. References: Park, 1946; Danner, 1965 and 1966.

Figs. 21-22. Lower Eocene (Ypresian) limestone in flysch sequence of Zumaya, Spain.

Globorotalia subbotinae–Globorotalia lensiformis zones. Locality: east side, Playa de San Telmo.

A thin, gray, marly limestone bed, sandy at base, passing upward into shale. Part of a flysch section containing many beds, each grading from (1) sandstone to shale, or (2) sandstone to limestone to shale, or (3) limestone to shale. The limestones contain abundant planktonic foraminifers, the shales and sandstones abundant ichnofossils. Reference: von Hillebrandt, 1965. (This author draws the Paleocene-Eocene boundary higher than most current workers, and considers these beds as Late Paleocene.)

Fig. 21. A coccosphere of *Coccolithus* sp., filled by a single calcite crystal and surrounded by a mosaic of variously derived calcite grains. Individual coccoliths and large single crystal show faint growth lines. Note the tight fit between the individual coccoliths in the coccosphere; cf. drawing, Fig. 5D. Note also the "marbled" pattern of dark spots on the larger calcite grains, which we are inclined to attribute to imperfections in the replica. Surfaces of these larger grains also show a granular "tea leaf" pattern. Photo Honjo.

SCALE BAR 5 MICRONS

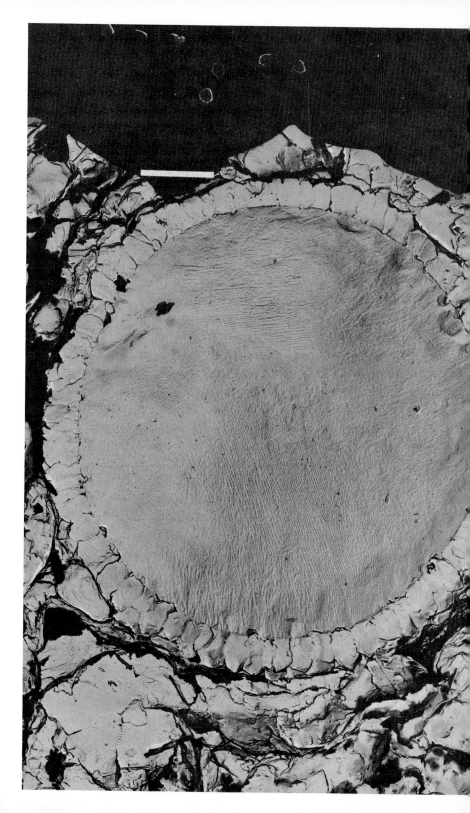

Fig. 22. *Thoracosphaera* cf. *deflandrei*, filled by a single crystal of calcite. Wall composed of pavement-block-like grains of calcite. Black patches in matrix are clay extractions. The surface of this preparation has more relief than most of our preparations and may be an etch-modified fracture surface. The peculiar striations on the central calcite crystal, forming locally intersecting patterns, are probably the cleavage directions, of which one is almost parallel to the surface. The black areas at the top are the margin of the supporting metal grid. Photo Honjo.

SCALE BAR 5 MICRONS

Figs. 23-36. Paleocene (Montian-Landenian) limestone in flysch sequence of Zumaya, Spain.

Zone of *Globorotalia pusilla pusilla*/*Globorotalia angulata*. Locality: western part of Playa de San Telmo.

This specimen is from a bed of fine-grained, red, nodular limestone, lacking megafossils, but containing abundant planktonic foraminifers, *Thoracosphaera*, and *Braarudosphaera*, all of which are visible with the optical microscope. Quartz silt is also present. Electron microscopy shows this rock to consist mainly of coccolith remains, sutured together by solution welding and local rim-cement overgrowths. References: Honjo and Fischer, 1964 and 1965a; von Hillebrandt, 1965.

· 55 ·

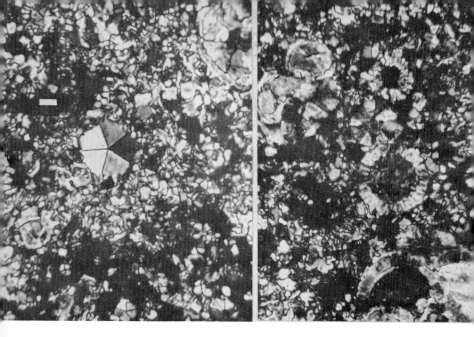

Fig. 23. Optical photomicrographs, 860×, of a very thin
section (<5 microns), in X-polarized light. Large coccoliths
and coccospheres in poorly resolved, vermicular-appearing
matrix, which is resolved in electron micrographs (Figs.
24-36) taken from same hand specimen. Left, a pentalith of
Braarudosphaera bigelowi, showing one of its five plates in
extinction position; cf. Fig. 24. Right, coccospheres of
Thoracosphaera saxea: In center, a transverse section, showing extinction cross; above, a more nearly tangential section,
showing interlocking nature of wall blocks; cf. Figs. 8, 27.
Between the thoracospheres, an obscure cross section of a
Braarudosphaera coccosphere; cf. Figs. 5, 31. At lower right
and upper middle, globigerinid Foraminifera. Photo Fischer.

SCALE BAR 5 MICRONS

Fig. 24. Coccolith limestone. Equatorial section of a
Coccolithus sp. in lower center, cross sections in many
places. *Braarudosphaera bigelowi* pentaliths in equatorial
section at upper center, in cross section at lower right. Note
embayed, solution-welded margins of fossils. Photo Honjo.

Fig. 25. Coccoliths, at higher magnification, showing solution welding. Photo Honjo. From Honjo and Fischer, 1965.

SCALE BAR 5 MICRONS

Fig. 26. Equatorial section of a *Coccolithus* sp., showing two circlets of imbricated plates spiralling in opposite directions. Each such plate is a crystal unit. Note concentric growth lines within plates. Photo Honjo. From Honjo and Fischer, 1964 and 1965.

Paleocene flysch, Zumaya, Spain.

Fig. 27. Coccolith limestone, with cross section of a *Thoracosphaera saxea.* Wall of calcisphere composed of large pavement-block grains, coarser than those of the Eocene *Thoracosphaera* illustrated in Fig. 22. Grains of wall show enlargement by rim-cementation on inner and outer surfaces. Center filled by a single, large calcite crystal showing several directions of striations, which may be in part cleavage, in part growth lines; note granular "tea leaf" pattern on this and adjacent crystals. Matrix of solution-welded coccolith remains, including some good cross sections of *Coccolithus* sp. (cf. Fig. 5E) . Photo Honjo.

SCALE BAR 5 MICRONS

Fig. 29. At top, a *Thoracosphaera* with complex wall structure and a multicrystalline filling. Matrix mainly of coccoliths, including various sections of *Braarudosphaera* pentaliths. Photo Honjo.

SCALE BAR 5 MICRONS

Fig. 28. Tangential section of *Thoracosphaera saxea*, showing slightly interlocked block structure. The amoeboid grain at lower right (*a*) is probably sparry calcite cement. The double plate to the right of this (*bb*) and the rosette of three plates to the left of it (*ccc*) are probably sections through *Braarudosphaera* pentaliths. Photo Honjo.

Paleocene flysch, Zumaya, Spain.

Fig. 30. Solution-welded coccolith limestone. At lower left, part of a globigerinid foraminifer, showing coarsely crystalline chamber fillings, in part syntaxial with the wall (cf. Fig. 5A). At lower right, fragment of a thin-walled *Thoracosphaera* showing axially perforated wall blocks. At upper left, a rod-shaped rhabdolith (*r*). Photo Honjo:

SCALE BAR 5 MICRONS

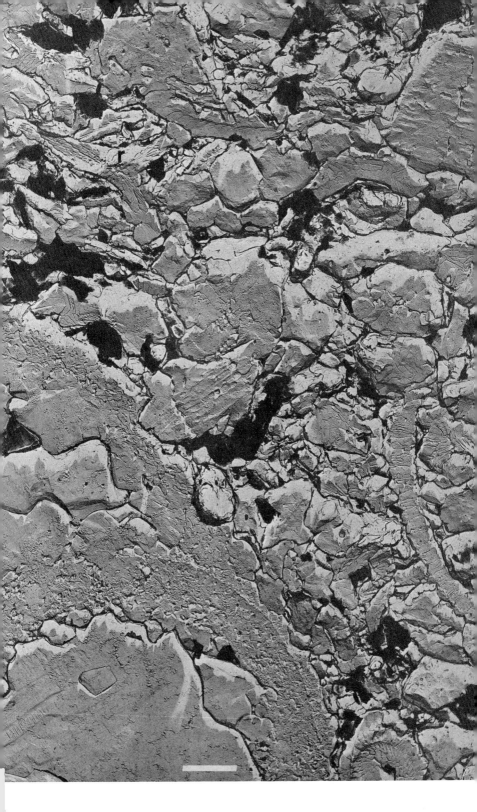

Paleocene flysch, Zumaya, Spain.

Fig. 31. A coccosphere of *Braarudosphaera*, showing typical thick pentaliths and the interior cavity filled by inward growth of calcite (syntaxial rim cement; cf. Figs. 5B and 5C). Photo Honjo.

SCALE BAR 5 MICRONS

Fig. 32. Tangential section of a coccosphere of *Braarudosphaera bigelowi*, centered on the boundary between two pentaliths, and cutting obliquely through parts of the other two pentaliths so as to reveal two segments of one, and three of the other (cf. Figs. 5B, 5C, 7). Abundant small coccoliths in the matrix. Photo Honjo.

Paleocene flysch, Zumaya, Spain.

Fig. 33. Tangential section of a coccosphere of *Braarudosphaera bigelowi,* centered on a pentalith and intersecting the edges of the five adjacent pentaliths. The pentaliths are thick, inward-tapering blocks of calcite, each consisting of five differently oriented crystal segments. The central pentalith has been intersected tangentially to its inner surface (cf. Figs. 5B, 5C, and 7). Note growth lamination of pentaliths. Photo Honjo.

SCALE BAR 5 MICRONS

Fig. 34. Mainly blade-shaped grains, most of which show a central groove—perhaps a twinning lamella. We suspect an organic origin for these grains. A few coccoliths are present as well. Photo Honjo.

SCALE BAR 5 MICRONS

Fig. 35. Parts of a globigerinid foraminifer, showing several chambers filled with calcite spar. The test wall shows the typical central groove and ridge formed by etching. The crystalline structure of the test wall appears to have been reorganized into blocky crystals, which have developed subhedral rim-cement extensions toward the exterior and interior (cf. Fig. 5A) . Note the "tea leaf" type granularity on the test wall. Black area at upper right corner is part of supporting metal grid. Coccoliths at lower right. Photo Honjo.

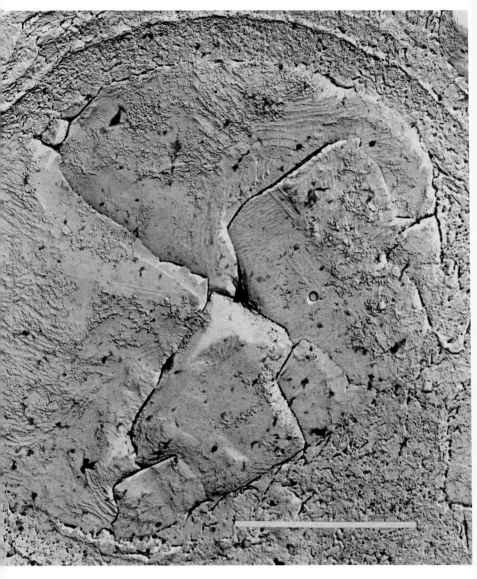

Fig. 36. Portion of a globigerinid chamber, showing a well-defined etching ridge in the middle of the wall. Chamber filled with a few relatively large calcite crystals. Photo Honjo.

SCALE BAR 5 MICRONS

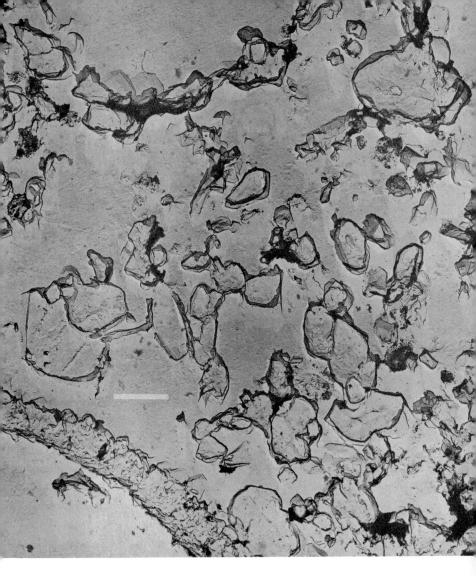

Cretaceous (Maestrichtian) chalk, Denmark.

Fig. 37. Preparation made from a plastic-impregnated
sample. The smooth "background" represents pore space
filled by plastic. Chalk comprised of subhedral calcite grains,
with scattered coccoliths and foraminiferal tests. Photo
Honjo.

Figs. 38-43. Cretaceous limestones from the Franciscan Formation, California.

The Franciscan Formation of western California is a Jurassic-Cretaceous, eugeosynclinal assemblage of sandstones, radiolarian cherts, pillow lavas, etc. Small amounts of fine-grained limestone are present as generally lens-shaped bodies, usually interbedded with volcanic rocks. Two general types of limestone are recognized by Bailey, Irwin, and Jones (1964) : (1) red limestones of the Laytonville type and (2) gray to white limestones of the Calera type. Both types contain planktonic foraminifers, by means of which they have been dated as Albian to Cenomanian. The red limestones are found mainly in areas north of San Francisco, the gray ones in areas to the south, particularly on the San Francisco Peninsula.

Electron microscopy also reveals striking differences between the two types of limestone. The red, Laytonville-type limestones contain abundant coccoliths and are, at most, only slightly recrystallized (Figs. 38-41). The gray, Calera-type limestones, in contrast, have been largely recrystallized to pavement mosaics (Figs. 42 and 43) ; however, they contain a few fragments of coccoliths and other problematic nannofossils, suggesting they may have once been coccolith limestones, perhaps thermally metamorphosed during Mid- to Late Cretaceous batholithic intrusion in the California Coast Ranges (Curtis *et al.*, 1958). References: Bailey, Irwin, and Jones, 1964; Garrison and Bailey, *in press*.

Fig. 38. Laytonville-type coccolith limestone, showing equatorial sections of several coccoliths. Specimen from Smith Ridge in Sonoma Co., California. Photo Garrison.

SCALE BAR 5 MICRONS

Fig. 39. Same specimen, showing equatorial section of a large, elongate coccolith and numerous coccolith fragments. Photo Garrison.

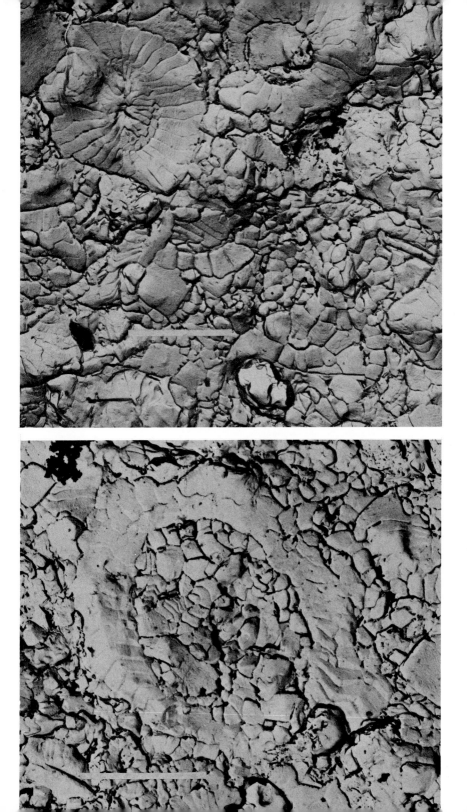

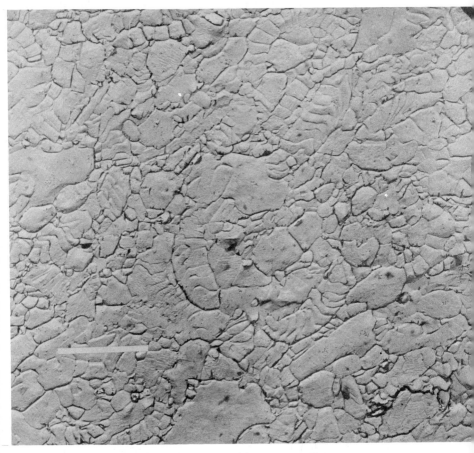

Fig. 40. Slightly more recrystallized coccolith limestone showing preferential grain elongation in "NE–SW" direction. Some of the transversely cut coccoliths are bent and broken, and part of the elongation appears to have resulted from solution welding. Photo Garrison.

SCALE BAR 5 MICRONS

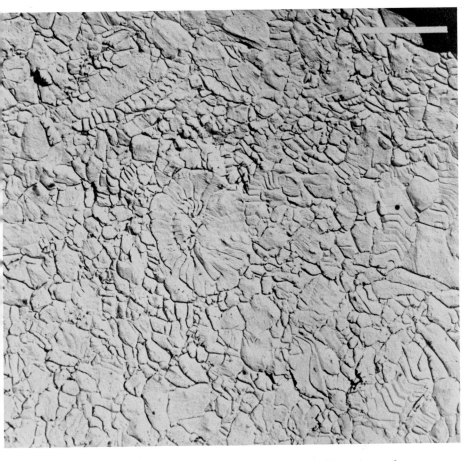

Fig. 41. Abundant coccoliths and some distortion of fabric. The equatorially cut *Coccolithus* specimen in the center is distorted, and the plates on its right side have lost their crystallographic identity. Photo Garrison.

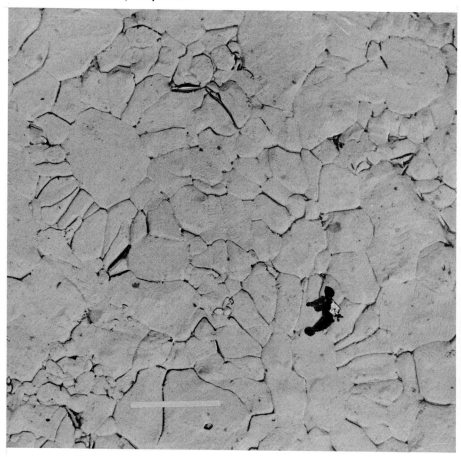

Fig. 42. Limestone from Black Mountain, on Monte Bello Ridge, Santa Clara Co. Coarser pavement mosaic, with relicts of three problematical nannofossils. Photo Garrison.

SCALE BAR 5 MICRONS

Fig. 43. Limestone from Cahil Ridge, San Mateo Co. Pavement mosaic of clear, relatively coarse grains, some of which reveal traces of sutures attesting their coccolith origin. Fabric suggests recrystallization under thermal stress. Photo Garrison.

Figs. 44-45. Upper Jurassic Solnhofen limestone, Solnhofen, Germany.

Perhaps the most famous of all limestones because of its unusual fossil content and the fact that this fine- and even-grained rock has long been used in lithography, in buildings, and in experimental work on rock deformation. References: Barthel, 1964; Flügel and Franz, 1967.

Fig. 44. Intricate "amoeboid" mosaic and abundance of inclusions contrast with the clean grains and "pavement" mosaic of the Franciscan limestones illustrated in Figs. 38-43. The granulated pattern on most of the grains, which in part somewhat resembles a "tea leaf" pattern, may be an artifact. Photo Honjo.

SCALE BAR 5 MICRONS

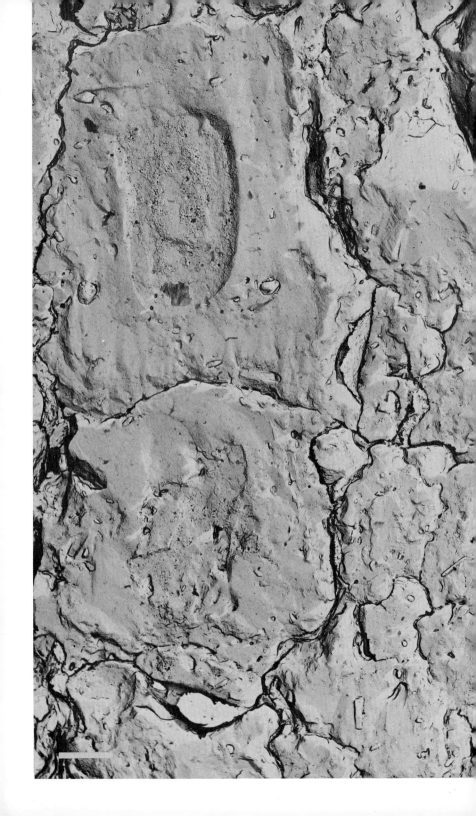

Fig. 45. Limestone showing amoeboid mosaic of grains with abundant inclusions, and showing at least two skeletal grains (large grains at left), possibly of echinoderm origin. Note presence of two types of inclusions: solid grains raised on original surface, and appearing as pits on photograph, without shadows; and fluid inclusions, raised grains on replica, throwing "shadows." Photo Honjo.

SCALE BAR 5 MICRONS

Lower Cretaceous (Valanginian?) Rossfeld Beds, Unken Valley, Northern Limestone Alps, Salzburg, Austria.

Fig. 46. A cross section of the large coccolith (?) *Nannoconus* sp. (Garrison, 1964, 1967). This *Nannoconus* shows many more prisms than do the Tithonian nannoconids of Figs. 59-60. Photo Garrison.

Figs. 47-64. Upper Jurassic to Lower Cretaceous lime-stones from the Oberalm Beds of the Unken Valley, Northern Limestone Alps, Austria.

Figs. 48-50 are Lower Cretaceous (Berriasian) and from same sample; Figs. 47 and 51-64 are Upper Jurassic (Tithonian).

The term Oberalm Beds is a local name for evenly inter-bedded drab limestones and shales of Late Jurassic–Early Cretaceous age; these rocks are widespread in the Alps, where they are commonly known as the *Aptychus* limestone in the north, and as Biancone or Majolica in the southern Alps and Apennines.

The Oberalm Beds are a mixture of beds of gray to buff, fine-grained, pelagic limestones, containing coccoliths, cal-pionellids and radiolarian tests, and allodapic limestones introduced by currents from adjacent shelf areas. Only the pelagic limestones are illustrated here. These rocks are much more tightly solution-welded than are the Paleocene limestones of the Spanish flysch, and the coccoliths have undergone more recrystallization, showing a tendency to lose the sutures between the plates. References: Colom, 1955; Garrison, 1964, 1967; Honjo and Fischer, 1964, and 1965a; Honjo, Fischer, and Garrison, 1965; Flügel and Fenninger, 1966.

Fig. 47. Optical photomicrograph, 560×, of a very thin section, showing several *Calpionella* tests and a calcitized and spar-filled radiolarian, in fine and poorly resolved matrix. Figs. 52, 54, 55, 57, and 60 are electron micrographs made from the same hand specimen. In these the matrix is re-solved and shows many coccolithophorid remains. Photo Fischer.

Fig. 48. Coccolith limestone with *Nannoconus steinmanni*. Some of the polygonal calcite grains are phantoms of coc-coliths, retaining traces of the characteristic zig-zag sutures, despite having apparently become converted to single calcite crystals. Photo Honjo.

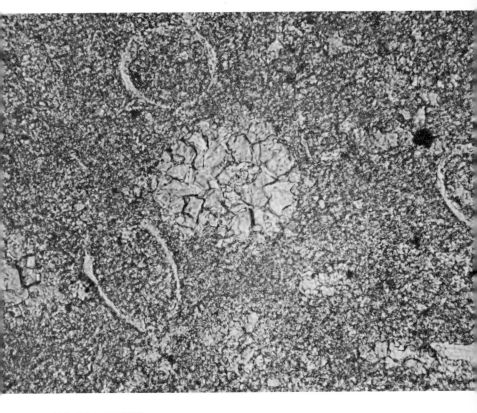

SCALE BAR 5 MICRONS

Fig. 49. Coccolith limestone. Note loss of distinctness in coccolith sutures and development of clean, unsutured grains which probably represent "microspar" calcite filling initial voids. The prominent grain in upper right is probably quartz. Photo Honjo.

Fig. 50. A tiny coccolith (?) engulfed by a large calcite crystal of secondary origin. Photo Honjo.

SCALE BAR 5 MICRONS

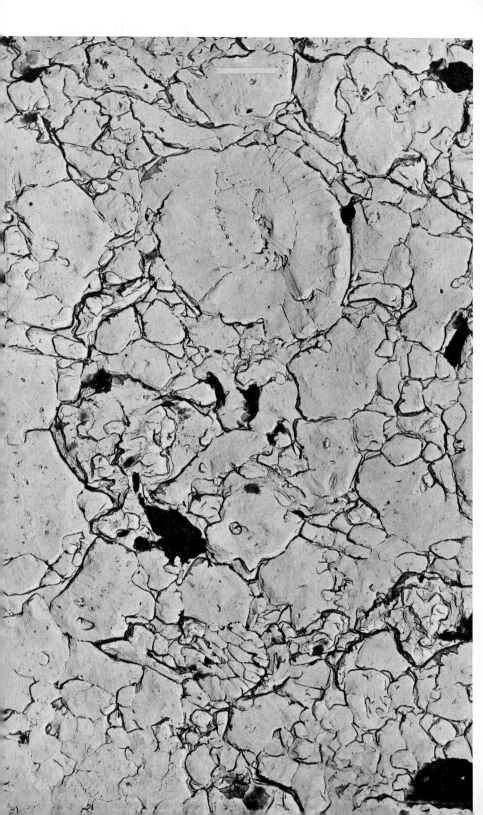

Upper Jurassic (Tithonian), Oberalm Beds, Unken Valley, Austria.

Fig. 51. A recrystallized coccolith limestone. Equatorially cut coccolith at upper right shows progressive loss of suture definition toward upper left. Some other grains show coccolith origin by retention of sutures of varying stages of distinctness. *Nannoconus* (group of prisms) at lower center. Photo Honjo.

SCALE BAR 5 MICRONS

Fig. 52. Calpionellid wall, showing etching ridge in mid-part of wall. Wall is composed of faintly defined calcite crystals, probably extended on inner and outer surfaces by rim cementation (cf. Fig. 9). In lower part of picture are remnants of *Nannoconus* (?) and coccoliths. At top center, a small coccolith extends deeply into the test, suggesting that it may have been built into the skeleton by the *Calpionella*. Photo Honjo.

Upper Jurassic (Tithonian) pelagic limestones of the Oberalm Beds, Unken Valley, Austria.

Fig. 53. A well-preserved *Coccolithus,* showing growth banding of the individual plates. Photo Garrison.

Fig. 54. Solution-welded fabric, showing remnant of *Coccolithus,* partly recrystallized, surrounded by mosaic of clean calcite grains. Photo Honjo. From Honjo and Fischer, 1964.

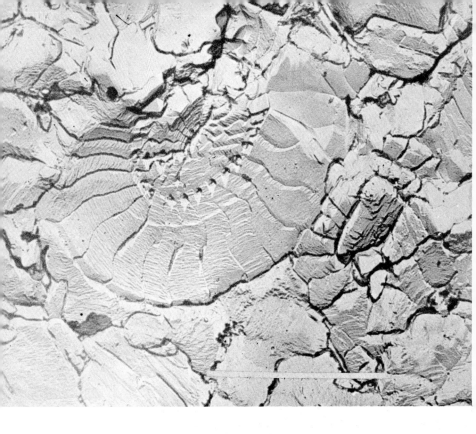

SCALE BAR 5 MICRONS

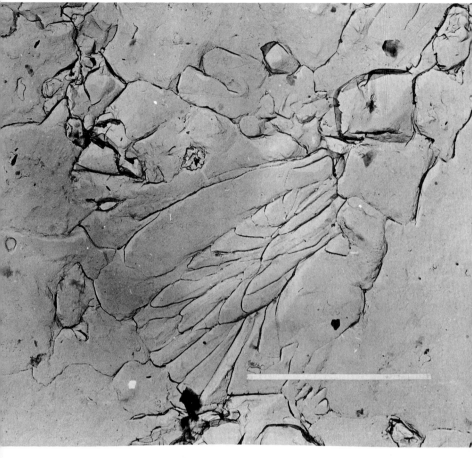

Fig. 55. Obliquely longitudinal section of a rhabdolith composed of two sets of predominantly radially set plates. Photo Honjo. From Honjo and Fischer, 1964.

Fig. 56. Cross section of a rhabdolith with two sets of predominantly radial plates. Photo Garrison.

Fig. 57. Cross section of a rhabdolith with more spirally arranged plates. Note narrow rim partly enveloping rhabdolith. Photo Honjo. From Honjo and Fischer, 1964 and 1965.

SCALE BAR 5 MICRONS

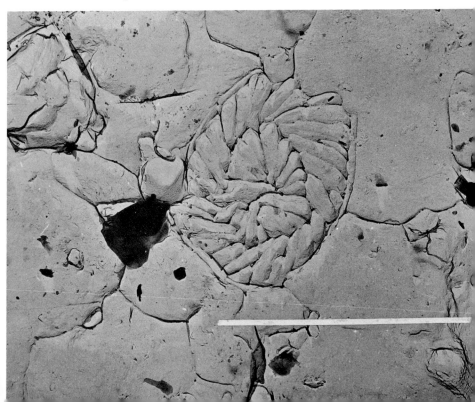

Upper Jurassic (Tithonian) Oberalm Beds, Unken Valley, Austria.

Fig. 58. *Nannoconus steinmanni* in longitudinal section. Photo Garrison. From Garrison, 1967.

Fig. 59. *Nannoconus steinmanni* in cross section; note loss of plate definition in lower left part of fossil. Same sample as Fig. 58. This *Nannoconus* shows fewer prisms than the Cretaceous one illustrated in Fig. 46, and more than the Tithonian one of Fig. 60. Photo Garrison.

SCALE BAR 5 MICRONS

Fig. 60. Cross sections of *Nannoconus* sp. showing few prisms. Photo Honjo. From Honjo and Fischer, 1964 and 1965.

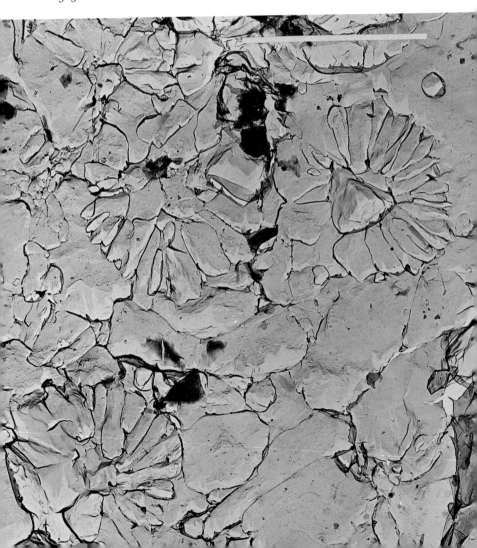

Upper Jurassic (Tithonian) Oberalm Beds, Unken Valley, Austria.

Fig. 61. Interlocking mosaic of comparatively coarse calcite grains, showing distinctive etch patterns and presumed twinning lamellae. Photo Honjo.

Fig. 62. Same. The smaller, prominent grains are probably of quartz. Photo Honjo.

SCALE BAR 5 MICRONS

Fig. 63. Coarse mosaic of calcite grains. Large grain at top shows growth zonation lines. Irregular clear line in center of picture is tear in replica and coincides with grain boundaries. Photo Honjo.

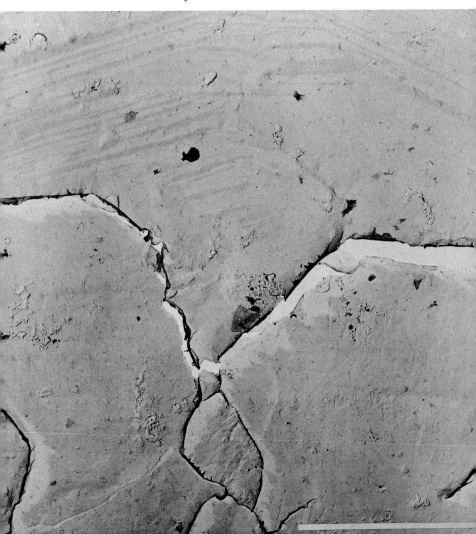

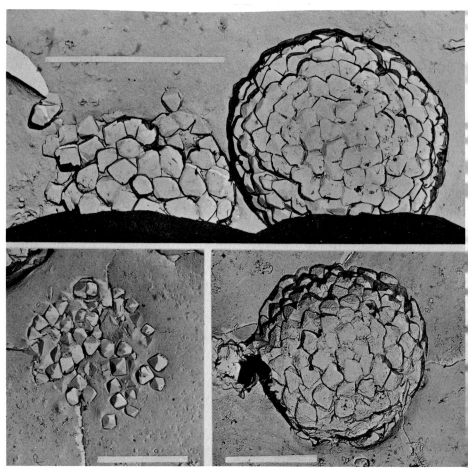

Fig. 64. Clusters of octahedral pyrite crystals.

Much of the pyrite in these beds occurs in sphaeroidal aggregates (framboids), lodged on the floors of radiolarian spheres (Honjo, Fischer, and Garrison, 1965). Polishing and etching has left the pyrite standing above the level of the surrounding calcite. Photo Honjo. From Honjo, Fischer, and Garrison, 1965.

SCALE BAR 5 MICRONS

Figs. 65-66. Middle to Upper Jurassic Ruhpolding Radiolarite, Unken Valley, Austria.

The term "radiolarite" applies to rocks with abundant radiolarian tests; such rocks may be radiolarian cherts or siliceous limestones, and they occur in Jurassic sections throughout the Alps. In the Unken Valley, the radiolarite section is mainly dense, red, siliceous limestone, with lesser amounts of radiolarian chert. These rocks, known locally as the Ruhpolding Radiolarite, are of Middle to Late Jurassic age, and lie beneath the Oberalm Beds. They are interpreted as having been siliceous and calcareous pelagic oozes.

Microscopically, the radiolarites are characterized by numerous poorly preserved radiolarian tests, and by intimate intermingling of silica, as microcrystalline quartz, with micritic calcite. The electron microscope reveals that the micrite is largely a recrystallization mosaic, but contains a few residual fragments of coccoliths. References: Grunau, 1959; Garrison, 1964.

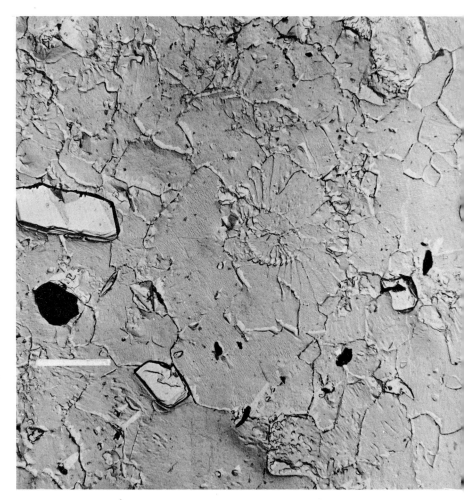

Fig. 65. Limestone showing mosaic of calcite grains and coccolith fragments; identity of rhomboidal grains unknown. Photo Garrison.

SCALE BAR 5 MICRONS

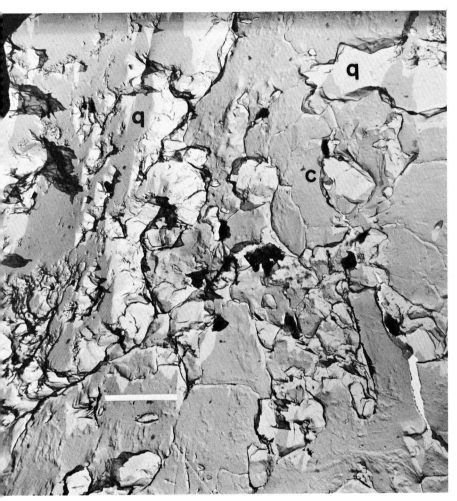

Fig. 66. Siliceous limestone showing grains with high relief (*q*), probably microcrystalline quartz, and micritic calcite (*c*). A coccolith fragment lies just to left of *c*. Photo Garrison.

Figs. 67-70. Jurassic (Sinemurian to Bajocian) Adnet Beds, Unken Valley, Austria.

The Adnet Beds lie below the Ruhpolding Radiolarite, and are of Early to Middle Jurassic age. They are mainly red, nodular limestones with numerous ammonites; such rocks in other parts of the Alps are known as "Ammonitico Rosso." At the top of the Adnet Beds are hard, red, fine-grained limestones, of probable Middle Jurassic (Bajocian) age, which contain manganese-iron oxide crusts and nodules. These limestones, from which the following illustrations were taken, contain abundant coccoliths, and probably represent pelagic oozes. Reference: Garrison, 1964.

Fig. 67. Equatorial section of a largely recrystallized coccolith in center, embayed by adjacent matrix grains. The large grain below it is also of coccolith origin, showing traces of diagnostic sutures. Matrix mosaic of clean, moderately amoeboid grains. Photo Garrison.

Fig. 68. A coccolith with distinctive arrangement of plates. Photo Garrison.

SCALE BAR 5 MICRONS

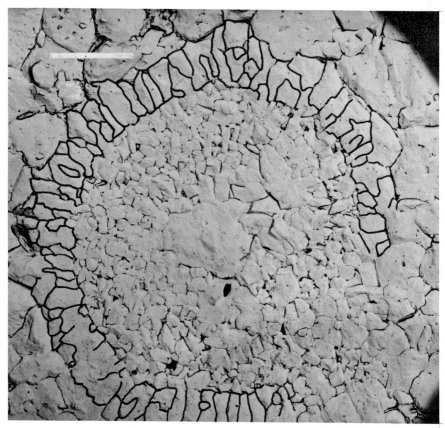

Fig. 69. *Thoracosphaera* sp., wall retouched, composed of prisms. Note large aperture at lower right. Interior of sphere partly filled with peculiarly subequal angular calcite prisms, apparently oblique cross sections of prisms similar to those forming the wall. Photo Garrison.

SCALE BAR 5 MICRONS

Fig. 70. *Thoracosphaera* sp., apparently an almost tangential section, providing oblique sections across the prisms which form the wall. Photo Garrison.

Figs. 71-75. Upper Triassic (Norian-Rhaetian) Dachstein limestone, Northern Limestone Alps, Austria.

The Dachstein limestone is a lagoonal facies between a barrier reef belt and the ultra-backreef 'Hauptdolomit (Zapfe, 1959; Fischer, 1966). Our figures illustrate several subfacies: Figs. 71 and 72 illustrate the somewhat dolomitic algal mat facies (loferite), of intertidal or supratidal origin, and show remains of algal filaments. Fig. 73 is of a pink, lutitic loferite without filaments. All three samples are from the Riemannhaus section (Fischer, 1966), in the Steinernes Meer range, above Saalfelden, Austria. Fig. 74 represents a sedimentary (neptunian) sill—a fine-grained, buff carbonate mud that was deposited within a crack (sheet crack) parallel to the bedding plane. The crack was presumably produced by desiccation and weathering of ˙sediment during a period of emergence, and the mud was introduced during the succeeding transgression. This mud in turn shrank and cracked parallel to bedding, its cracks becoming filled with radiaxial calcite to form a zebra rock (Fischer, 1966). Locality: Steinernes Meer, above Saalfelden, Austria. Fig. 75 illustrates a neptunian dike of red limestone cutting through Dachstein sediments perpendicular to bedding (Fischer, 1966). Locality: Steinernes Meer, above Saalfelden, Austria.

Fig. 71. Section transverse to algal filaments, which are filled with calcite spar; matrix composed predominantly of etch-resistant, fine-grained, anhedral dolomite. Photo Garrison.

Fig. 72. Section parallel to an algal filament. Matrix is chiefly fine-grained calcite, with clusters of fine anhedral dolomite grains, concentrated especially on the surface of the algal filament, and standing in relief above the calcite. Filament filled with sparry calcite. By analogy with the Persian Gulf Sabkhas and the Caribbean intratidal-supratidal areas, the dolomitization is attributed to an early diagenetic replacement of aragonite. Cf. Fig. 3C. Photo Garrison.

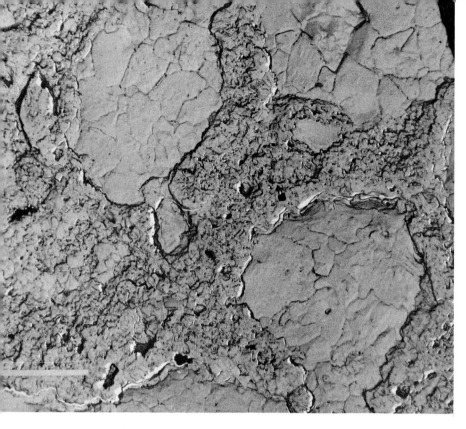

SCALE BAR 5 MICRONS

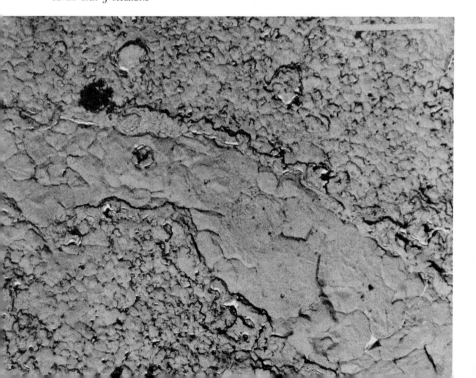

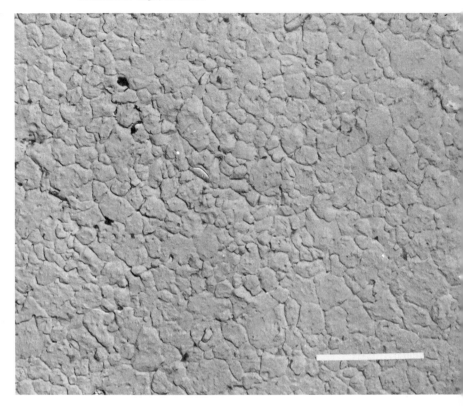

Fig. 73. Fine-grained, pink, calcilutite phase of a loferite, attributed to intratidal or supratidal origin. Note extremely fine grain. Photo Garrison.

Fig. 74. Internally deposited (sheet-crack filling), buff-colored calcilutite. Mosaic intermediate between amoeboid and pavement types. Photo Garrison.

Fig. 75. Neptunian dike of deep red, fine-grained lime-stone, with fairly coarse calcite grains. Some of the inclusions may be hematite. Cf. Fig. 2D. Photo Garrison.

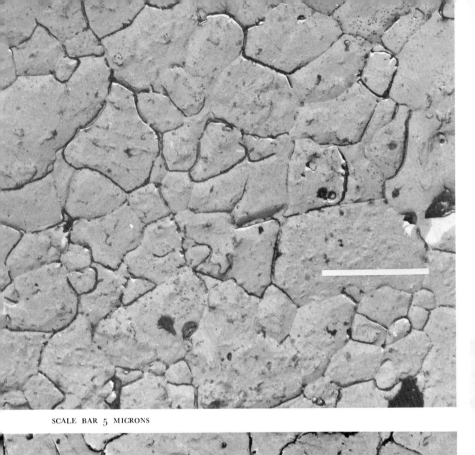

SCALE BAR 5 MICRONS

Figs. 76-82. Upper Triassic limestones of the Hallstatt "facies," Northern Limestone Alps, Austria and Germany.

The Hallstatt rocks include a wide variety of sedimentary rock types, which apparently accumulated within a starved basin in front of the barrier reefs (Fischer, 1966). Among these types are red, cephalopod-rich and manganiferous limestones such as the Carnian limestones of the Feuerkogel near Aussee, Austria (Figs. 76-78) ; the buff, cherty, nodular limestones at Pötschen Pass, between Ischl and Aussee, Austria (Figs. 79-81) ; and the shaly, black Pedata Beds at Pötschen Pass, which contain graded limestones representing reef-derived turbidites—of which Fig. 82 shows the finest phase.

Figs. 76 and 77. Red Carnian-Norian Hallstatt limestone from Feuerkogel. Amoeboid recrystallization mosaics retaining vestiges of round objects in the size range of coccospheres (cf. Fig. 3B). Photos Garrison.

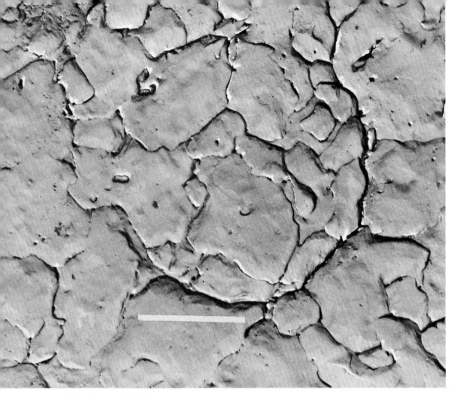

SCALE BAR 5 MICRONS

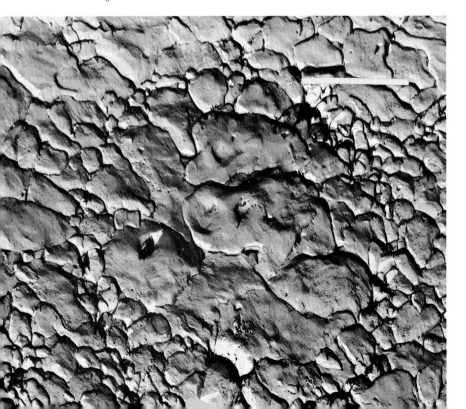

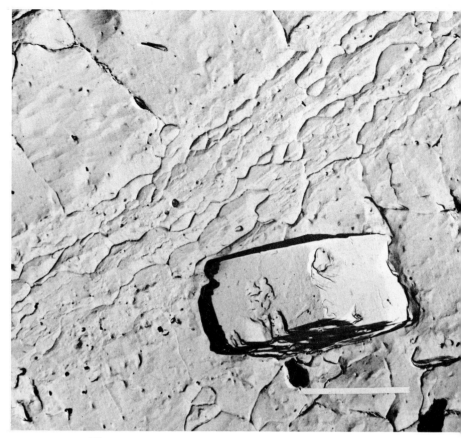

Fig. 78. Nodular, red Norian limestone from the Kälberstein quarry at Berchtesgaden. Picture is diagonally crossed by a thin shell—probably a somewhat recrystallized cephalopod shell (cf. Fig. 84), the hollow interior of which, at upper left, was filled with blocky calcite spar. Photo Garrison.

Figs. 79-80. Norian Pötschen limestone of the Hallstatt "facies," containing problematic aggregates of plates. We are uncertain whether these represent aggregates of platy minerals or whether they are calcareous skeletal structures. Fig. 80 shows also an elliptical fossil relict, outlined by grain boundaries and by small grains, in the size-range of coccoliths. Photos Garrison.

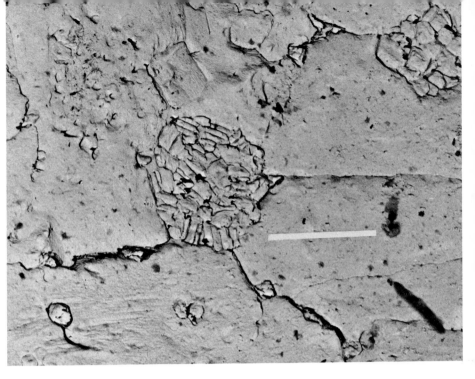

SCALE BAR 5 MICRONS

Fig. 81. Pötschen limestone, showing a larger aggregate of plates (cf. Figs. 76, 77) which suggests a solid grain composed of randomly oriented packages of plates (crystals?). The small dark grains between larger calcite grains, also present in Fig. 80, are problematic. Photo Garrison.

SCALE BAR 5 MICRONS

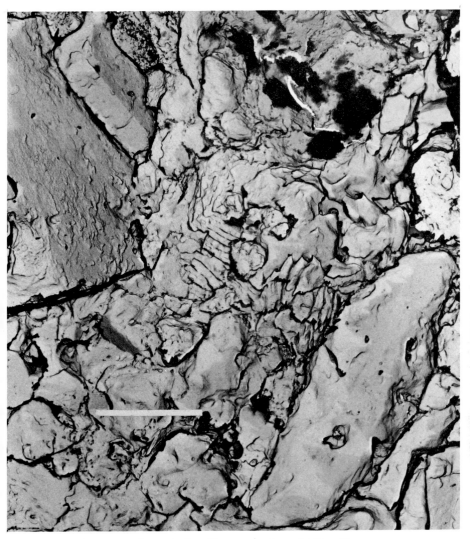

Fig. 82. Pedata Beds: Fine-grained turbidite limestone, showing solution-welded mixture of heterogeneous carbonate grains. Package of parallel plates in center suggest the round aggregates of Figs. 76-78, and may represent either a type of skeletal material or a platy, non-carbonate mineral. Photo Garrison.

Upper Triassic limestone from the Marble Bay Formation, S. Beale Black-Rock Quarry, Texada Island, British Columbia, Canada.

Fig. 83. A comparatively coarse, blocky mosaic of clean calcite grains suggesting thermal recrystallization (cf. Fig. 4D). The latter appears very likely because limestones of the Marble Bay Formation are intruded by stocks of diorite-gabbro and quartz-diorite, and by numerous basic dikes (Mathews and McCammon, 1957). References: McConnell, 1914; Mathews and McCammon, 1957. Photo Garrison.

SCALE BAR 5 MICRONS

Permian (Leonardian) Bone Spring Formation, Delaware Basin, West Texas.

Fig. 84. Black, bituminous laminated limestone, deposited under euxinic conditions (Newell *et al.*, 1953). Segment of a paper-thin shell, retaining its wall structure, which appears to be the "brick-and-mortar" structure of aragonitic mother-of-pearl (Grégoire, 1957). We take this to be an unaltered shell of a juvenile cephalopod. The small black grains throwing long shadows appear to be extracted mineral grains, possibly pyrite. The large clear grains standing out in relief are probably quartz; the dark membranes around them are probably relicts of the first replica (acetate peel). Photo Honjo.

SCALE BAR 5 MICRONS

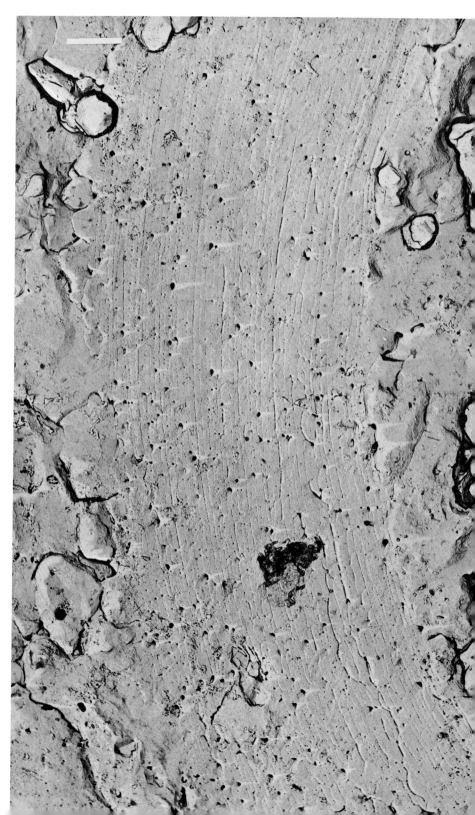

Figs. 85-89. Devonian Waterways Formation, Judy Creek Oil Field, Alberta, Canada.
A dark, bituminous limestone forming a deeper-water facies in front of the reefs (Murray, 1965, 1966). Photo Honjo.

Fig. 85. Limestone showing bimodal grain-size frequency. Are the patches of smaller grains relicts of a former uniformly fine-grained rock? Photo Honjo.

SCALE BAR 5 MICRONS

Devonian Waterways Formation, Judy Creek Oil Field, Alberta, Canada.

Fig. 86. Part of a megafossil, apparently a tubular structure in slightly oblique tangential section. Photo Honjo.

SCALE BAR 5 MICRONS

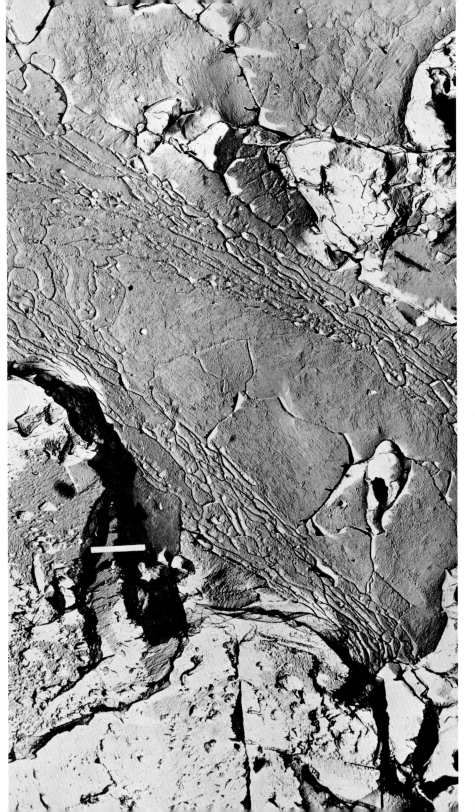

Devonian Waterways Formation, Judy Creek Oil Field, Alberta, Canada.

Fig. 87. Monocrystalline grains of calcite (light colored, generally rounded, but locally showing subhedral margins), and rhombs of inclusion-rich dolomite, in a vermicular matrix consisting of organic matter and possibly some clay (Murray, 1965, 1966). Black cloudy grains are probably clay extractions. The "marbled" pattern on the calcite grains is probably an artifact.

The irregularly scalloped margins of calcite grains are puzzling to us, but may have resulted from crystallization of calcite in this peculiar matrix. Photo Honjo.

SCALE BAR 5 MICRONS

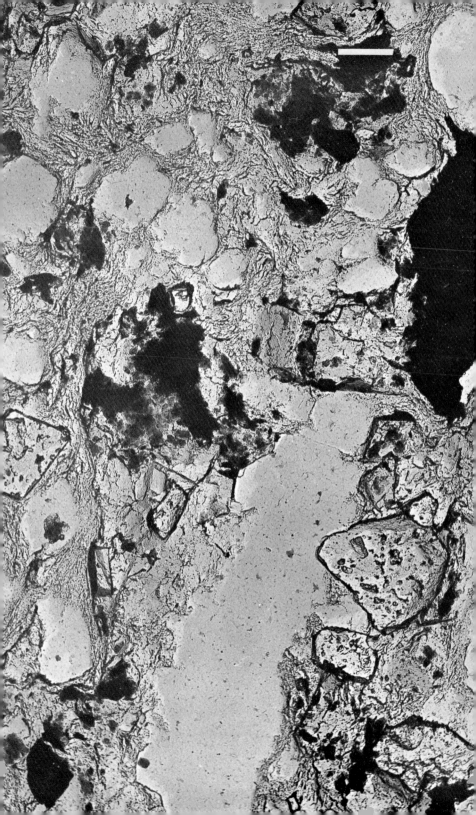

Devonian Waterways Formation, Judy Creek Oil Field, Alberta, Canada.

Fig. 88. Same as Fig. 87. Photo Honjo.

SCALE BAR 5 MICRONS

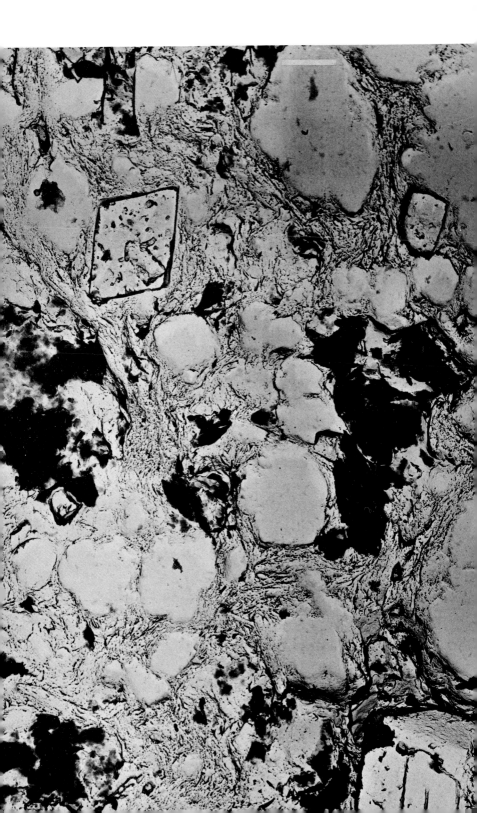

Devonian Waterways Formation, Judy Creek Oil Field, Alberta, Canada.

Fig. 89. Dolomite crystal (rounded rhombohedron) shows zoning expressed in sharp growth lines and in zonal distribution of holes (fluid inclusions). Photo Honjo.

SCALE BAR 5 MICRONS

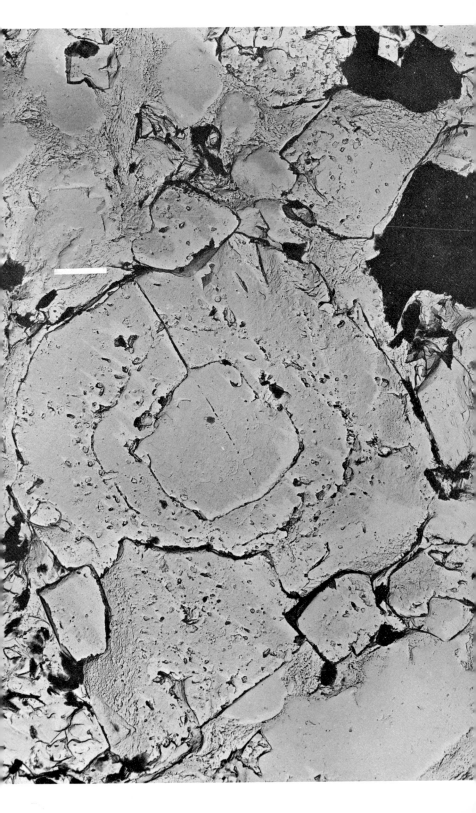

Lower-Middle Ordovician Marathon limestone, Marathon fold belt, Texas.

Fig. 90. A buff, graptolite-bearing limestone. A rather deeply etched section, showing elongate, intricately sutured calcite grains full of fluid inclusions. Photo Honjo.

SCALE BAR 5 MICRONS

Lower-Middle Ordovician Marathon limestone, Marathon fold belt, Texas.

Fig. 91. A deformed specimen, which shows development of a strongly directional fabric. with subhedral crystal terminations pointing in direction of elongation. Note the abundant slip and/or twinning planes, deeply etched, developed in some grains. Cf. Fig. 4C. Photo Honjo.

SCALE BAR 5 MICRONS

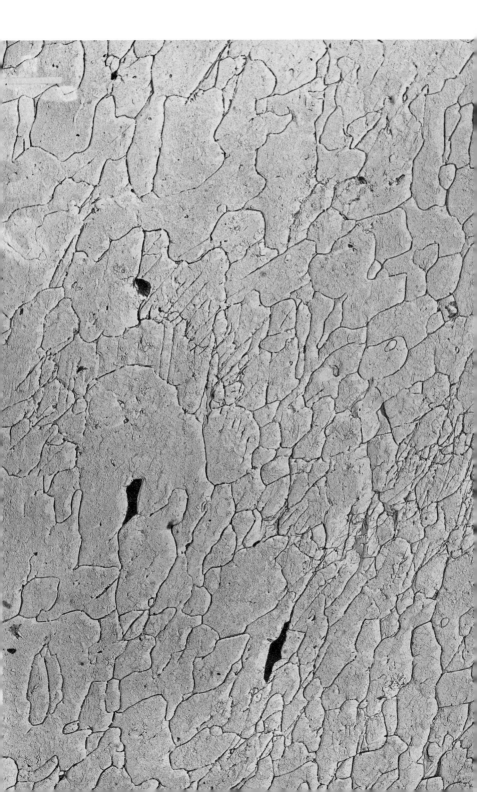

A black, fine-grained, pyritic limestone in the folded Appalachian Ridge-and-Valley belt. The calcite grains are relatively free from inclusions. Their shape is irregular, but not amoeboid, the boundaries taking the form of straight zig-zags.

Fig. 92. Mixture of coarser and finer calcite grains, and two rhomboidal dolomite grains. Photo Garrison.

Fig. 93. More uniform, slight coarser, calcite mosaic, with an unidentified grain in center. Photo Garrison.

**Lower Cambrian red limestone,
Avalon Peninsula, Newfoundland.**

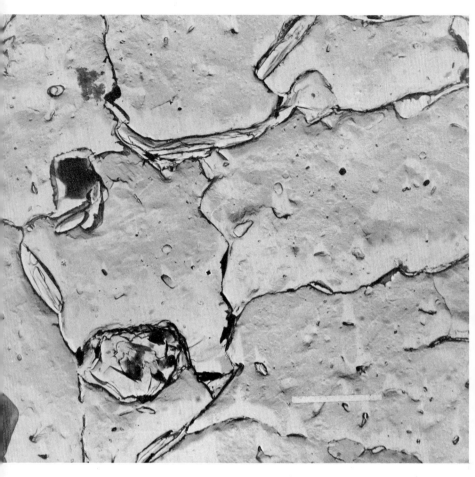

Fig. 94. Clay minerals or micas between the calcite grains.
Photo Garrison.

REFERENCES

Bailey, E. H., W. P. Irwin and D. L. Jones, 1964, Franciscan and related rocks, and their significance in the geology of western California: *Calif. Div. Mines Geol., Bull.* 183, 177 pp.

Barthel, K. W., 1964, Zur Enstehung der Solnhofener Plattenkalke (unteres Untertithon) : *Mitt. Bayer. Staatssamml. Paläontol. hist. Geol.,* v. 4, pp. 37-69.

Bathurst, R.G.C., 1958, Diagenetic fabrics in some British Dinantian limestones: *Liverpool Manchester Geol. J.,* v. 2, pp. 11-36.

Bathurst, R.G.C., 1964, The replacement of aragonite by calcite in the molluscan shell wall, pp. 357-379 in *Approaches to Paleoecology,* Imbrie and Newell, eds.: New York, John Wiley & Sons, Inc. 432 pp.

Baxter, J. W., 1960, *Calcisphaera* from the Salem (Mississippian) limestone in Southwestern Illinois: *J. Paleontol.,* v. 34, pp. 1153-1157.

Bonet, F., 1956, Zonificación mikrofaunistica de las calizas cretácicas del este de México: *Bol. Asoc. Mexicana Geólogos Petroleros,* v. 8, pp. 3-102.

Bradley, D. E., 1961, Replica and shadowing techniques, pp. 82-137 in *Techniques for Electron Microscopy,* Kay, ed.: Oxford, Blackwell, 331 pp.

Bramlette, M. N., 1958, Significance of coccolithophorids in calcium carbonate deposition: *Geol. Soc. Am. Bull.,* v. 69, pp. 121-126.

Bramlette, M. N. and E. Martini, 1964, The great change in calcareous nannoplankton fossils between the Maestrichtian and Danian: *Micropaleontology,* v. 10, pp. 291-322.

Colom, G., 1955, Jurassic-Cretaceous pelagic sediments of the Western Mediterranean zone and Atlantic area: *Micropaleontology,* v. 1, pp. 109-124.

Colom, G., 1956, Lito-facies y micropaleontologia de las formaciones Jurássico-neocomienses de la Sierra de Ricote

(Murcia) : *Bol. Inst. Geol. Minero España*, v. 67, pp. 13-15.

Curtis, G. H., J. F. Evernden and J. E. Lipson, 1958, Age determination of some granitic rocks in California by the Potassium-argon method: *Calif. Div. Mines Geol. Spec. Rept.* 54, 16 pp.

d'Albissin, M., 1959, Une application du microscope électronique à l'étude de la déformation des calcaires: *Rev. Géographie Phys. Géol. Dyn.*, v. 2, pp. 35-38.

Danner, W. R., 1965, Limestones of the western Cordilleran eugeosyncline of southwestern British Columbia, western Washington, and northern Oregon: *Wadia Commemorative Volume*, Min. Met. Inst. India, pp. 113-125.

Danner, W. R., 1966, Limestone resources of western Washington: *Washington Div. Mines Geol. Bull.* no. 52, 474 pp.

Deflandre, G., 1952, Sous-embranchement des Flagelles, pp. 107-115 in Piveteau, J., *Traité de Paléontologie*, vol. 1: Paris, Masson & Cie.

Doben, K., 1963, Über Calpionelliden an der Jura/Kreide Grenze: *Mitt. Bayer. Staatssamml. Paläontol. hist. Geol.*, v. 3, pp. 35-50.

Farinacci, A., 1964, Microorganismi dei Calcari "Maiolica" e "Scaglia" osservati al microscopio elettronico (Nannoconi e Coccolithophoridi) : *Boll. Soc. Paleontol. Italiana*, v. 3, pp. 172-181.

Fischer, A. G., 1966, The Lofer cyclothems of the alpine Triassic, pp. 107-149 in "Symposium on Cyclic Sedimentation," Merriam, ed.: *Kansas Geol. Surv. Bull.* 169, 636 pp.

Fischer, A. G., and R. E. Garrison, 1967, Carbonate lithification on the sea-floor: *J. Geol. (in press)*.

Flügel, E., 1967, Elektronenmikroskopische Untersuchungen an mikritischen Kalken. *Geol. Rundschau*, v. 56, pp. 341-358.

Flügel, E., and H. E. Franz, 1967, Elektronenmikroskopischer Nachweis von Coccolithen im Solnhofener Plattenkalk (Ober-Jura). *Neues Jahrb. Geol. Paläontol. Abhandl.*, v. 124, pp. 245-263.

REFERENCES

Flügel, H., and A. Fenninger, 1966, Die Lithogenese der Oberalmer Schichten und der mikritischen Plassen-kalke (Tithonium, Nördliche Kalkalpen) : *Neues Jahrb. Geol. Paläontol. Abhandl.*, v. 123, pp. 249-280.

Folk, R. L., and C. E. Weaver, 1952, A study of the texture and composition of chert: *Am. J. Sci.*, v. 250, pp. 498-510.

Garrison, R. E., 1964, Jurassic and Early Cretaceous sedimentation in the Unken Valley area, Austria: Unpublished Ph.D. dissertation, Princeton Univ., 188 pp.

Garrison, R. E., 1967, Pelagic limestones of the Oberalm Beds (Upper Jurassic-Lower Cretaceous), Austrian Alps: *Bull. Can. Petroleum Geology*, v. 15, pp. 21-49.

Garrison, R. E., and E. H. Bailey, *in press*, Electron microscopy of Franciscan limestones in California: *U.S. Geol. Surv. Res. Prof. Paper.*

Grégoire, C., 1957, Topography of the organic components in mother-of-pearl: *J. Biophys. Biochem. Cytol.*, v. 3, p. 797.

Grégoire, C., 1959, A study of the remains of organic components in fossil mother-of-pearl. *Inst. Roy. Sci. Nat. Belg. Bull.*, v. 35, no. 13, pp. 1-13.

Grégoire, C., and C. Monty, 1963, Observations au microscope électronique sur le calcaire à pate fine entrant le constitution des structures stromatolithiques du Viseen Moyen de la Belgique: *Geol. Soc. Belg., Ann.*, no. 10, pp. 389-397.

Grunau, H. R., 1959, *Mikrofazies und Schichtung ausgewählter jungmesozoischer, Radiolarite-führender Sedimentserien der Zentral-Alpen*: Leiden, E. J. Brill, 179 pp.

Grunau, H. R., and H. Studer, 1956, Elektronenmikroskopische Untersuchungen an Bianconekalken des Südtessins: *Experientia*, v. 12, pp. 141-150.

Harvey, R. D., 1966, Electron miscroscope study of microtexture and grain surfaces in limestones: *Illinois State Geol. Surv.*, Circ. 404, 18 pp.

Hay, W. W., and K. M. Towe, 1962, Electron-microscope studies of *Braarudosphaera bigelowi* and some related Coccolithophorids: *Science*, v. 137, pp. 426-428.

REFERENCES

Hillebrandt, A. von, 1965, Foraminiferen-Stratigraphie im Alttertiär von Zumaya (Provinz Guipuzcoa, NW-Spanien (und ein Vergleich mit anderen Tethys-Gebieten) : *Bayer. Akad. Wiss., Math.-Nat. Kl.*, N.F., Heft 123, 62 pp.

Honjo, S., and A. G. Fischer, 1964, Fossil coccoliths in limestone examined by electron miscroscopy: *Science*, v. 144, pp. 837-839.

Honjo, S., and A. G. Fischer, 1965a, Paleontological investigations of limestones by electron miscroscopy, pp. 326-334 in *Handbook of Paleontological Techniques*, Kummel and Raup, eds.: San Francisco, W. H. Freeman & Co., 852 pp.

Honjo, S., and A. G. Fischer, 1965b, Thin sections and peels for high-magnification study and phase-contrast microscopy, pp. 241-247 in *Handbook of Paleontological Techniques*, Kummel and Raup, eds.: San Francisco, W. H. Freeman and Co., 852 pp.

Honjo, S., A. G. Fischer and R. E. Garrison, 1965, Geopetal pyrite in fine-grained limestones: *J. Sediment. Petrology*, v. 35, pp. 480-488.

Illing, L. V., A. J. Wells and C. M. Taylor, 1965, Penecontemporaneous dolomite in the Persian Gulf, pp. 89-111, in "Dolomitization and Limestone Diagenesis," Pray and Murray, eds.: *Soc. Econ. Paleontol. Mineral. Spec. Publ.* no. 13, 180 pp.

Kamptner, E., 1927, Beiträge zur Kenntnis adriatischer Coccolithophoriden. *Arch. Protistenk.*, v. 38, pp. 173-184.

Kamptner, E., 1931, *Nannoconus steinmanni* n.g.n.s., ein merkwürdiges gesteinsbildendes Mikrofossil aus dem jüngeren Mesozoikum der Alpen: *Paläontol. Z.*, v. 13, pp. 288-298.

Kamptner, E., 1946, Zur Kenntnis der Coccolithen-Gattung *Thoracosphaera* Kpt.: *Akad. Wissensch. Wien, Math.-Naturw. Kl., Anz.*, v. 11, pp. 100-103.

Mathews, W. H., and J. W. McCammon, 1957, Calcareous deposits of southwestern British Columbia: *Brit. Columbia Dept. Mines, Bull.* no. 40, 105 pp.

REFERENCES

McConnell, R. G., 1914, Texada Island, B. C.: *Geol. Surv. Canada, Mem.* 58, 112 pp.

Murray, J. W., 1965, Stratigraphy and carbonate petrology of the Waterways Formation, Judy Creek, Alberta, Canada: *Bull. Can. Petroleum Geology*, v. 13, pp. 303-326.

Murray, J. W., 1966, An oil producing reef-fringed carbonate bank in the Upper Devonian Swan Hills Member, Judy Creek, Alberta: *Bull. Can. Petroleum Geology*, v. 14, pp. 1-103.

Newell, N. D., J. K. Rigby, A. G. Fischer, A. J. Whiteman, J. E. Hickox and J. S. Bradley, 1953, *The Permian Reef complex of the Guadalupe Mountain Region, Texas and New Mexico*: San Francisco, Freeman and Co., 236 pp.

Noël, D., 1958, Étude de coccolithes du Jurassique et du Crétace inférieur. *Publ. Serv. Carte Géol. Algérie, Bull.* N.S., no. 20, pp. 157-195.

Park, C. F., 1946, The spilite and manganese problems of the Olympic Peninsula, Washington: *Am. J. Sci.*, v. 244, pp. 305-323.

Rupp, A. W., 1966, Origin, structure and environmental significance of Recent and fossil calcispheres: *Geol. Soc. Am., Ann. Meeting* (abstract).

Seeliger, R., 1956, Übermikroskopische Darstellung dichter Gesteine mit Hilfe von Oberflächenabdrücken: *Geol. Rundschau*, v. 45, pp. 332-336.

Shinn, E. A., R. N. Ginsburg and R. M. Lloyd, 1965, Recent supratidal dolomite from Andros Island, Bahamas, pp. 112-123, in "Dolomitization and Limestone Diagenesis," Pray and Murray, eds.: *Soc. Econ. Paleontol. Mineral. Spec. Publ.* no. 13, 180 pp.

Shoji, R., and R. L. Folk, 1964, Surface morphology of some limestone types as revealed by electron microscope: *J. Sed. Petrology*, v. 34, pp. 144-155.

Stradner, H., 1961, Vorkommen von Nannofossilien im Mesozoikum und Alttertiär: *Erdoel*, v. 77, pp. 77-88.

Zapfe, H., 1959, Faziesfragen des alpinen Mesozoikums: *Geol. Bundesanstalt, Wien, Verhandl.*, pp. 122-128.